Innovations in Enhanced and Improved Oil Recovery - New Advances

Edited by Mansoor Zoveidavianpoor

Published in London, United Kingdom

Innovations in Enhanced and Improved Oil Recovery - New Advances
http://dx.doi.org/10.5772/intechopen.111078
Edited by Mansoor Zoveidavianpoor

Contributors
Ahmad Khanifar, Ahmed Merzoug, Aimen Laalam, Arezou Jafari, Habib Ouadi, John Harju, Lynn Helms, Magda Ibrahim Youssif, Mansoor Zoveidavianpoor, Maolei Cui, Nafiseh Mehrooz, Naser Asadzadeh, Olusegun Stanley Tomomewo, Qiaoyu Ge, Reza Gharibshahi, Rui Wang, Zengmin Lun

First published in London, United Kingdom, 2024 by IntechOpen
IntechOpen is the global imprint of INTECHOPEN LIMITED, registered in England and Wales, registration number: 11086078, 5 Princes Gate Court, London, SW7 2QJ, United Kingdom

British Library Cataloguing-in-Publication Data
A catalogue record for this book is available from the British Library

Additional hard and PDF copies can be obtained from orders@intechopen.com

Innovations in Enhanced and Improved Oil Recovery - New Advances
Edited by Mansoor Zoveidavianpoor
p. cm.
Print ISBN 978-0-85466-197-8
Online ISBN 978-0-85466-196-1
eBook (PDF) ISBN 978-0-85466-198-5

For EU product safety concerns:
IN TECH d.o.o., Prolaz Marije Krucifikse Kozulić 3, 51000 Rijeka, Croatia,
info@intechopen.com or visit our website at intechopen.com.

Meet the editor

Dr. Mansoor Zoveidavianpoor is a distinguished professional with more than 24 years of versatile experience in both industry and academia, leading more than twenty-five projects that underscore his exceptional leadership in steering initiatives to successful completion. His extensive journey encompasses various technical, operational, and managerial roles, contributing significantly to the fields of petroleum engineering and energy transition. Armed with a bachelor's degree in Geology and further academic prowess, Dr. Zoveidavianpoor holds both a master's and a Ph.D. in Petroleum Engineering. Throughout his career, he has demonstrated expertise in an array of disciplines, ranging from project management and flow assurance to drilling, production engineering, reservoir management, and Carbon Capture and Storage (CCS). Dr. Zoveidavianpoor has shared his wealth of knowledge through lecturing at prestigious institutions, including the University of Technology Malaysia (UTM) and the Petroleum University of Technology (PUT), and Islamic Azad University (IAU), Iran. His commitment to education aligns with his passion for fostering the next generation of professionals in the energy industry. As a member of the Society of Petroleum Engineers (SPE) and a Chartered Petroleum Engineer registered at the Energy Institute, Dr. Zoveidavianpoor exemplifies a commitment to professional excellence. His recognition as an Environmental Impact Assessment (EIA) subject specialist at the Department of Environment (DOE), Malaysia, reflects his dedication to sustainable practices in the energy sector. Dr. Zoveidavianpoor's scholarly contributions extend beyond the classroom. With more than fifty publications in international peer-reviewed journals and conferences, along with his involvement in five textbooks, he has actively shaped the discourse in the field. His participation in numerous scientific committees showcases his commitment to advancing research and knowledge in petroleum engineering. Leading more than twenty-five projects, publishing sixty papers in indexed journals and conferences, editing five books, holding two patents, three Technology Readiness Levels (TRLs), and pioneering technologies highlight Dr. Zoveidavianpoor's trailblazing impact in pushing the boundaries of innovation in the dynamic field of petroleum engineering. Currently an associate professor at UTM, Dr. Zoveidavianpoor continues to play a pivotal role in shaping the academic landscape. Simultaneously, he remains engaged in several consulting projects, leveraging his wealth of experience in petroleum engineering and contributing to ongoing energy transition initiatives. Dr. Zoveidavianpoor's multidisciplinary approach and current focus on energy transition underscore his commitment to addressing the complex challenges facing the energy industry. His unwavering dedication to excellence, combined with his extensive expertise, positions him as a key figure driving innovation and progress in the dynamic field of petroleum engineering.

Contents

Preface

I am delighted to present this compilation of cutting-edge research and advancements in enhanced and improved oil recovery, encapsulated in the book *Innovations in Enhanced and Improved Oil Recovery – New Advances*. As the editor of this volume, it is my privilege to introduce the diverse array of chapters contributed by experts in the industry, each offering unique insights and perspectives.

Before delving into the details of the chapters, I extend my sincere appreciation to the contributors for their dedication and invaluable contributions. Their expertise has enriched this volume, providing a comprehensive overview of the latest innovations in enhanced and improved oil recovery.

The journey through this book begins with Chapter 1, "Introductory Chapter: Lithological Dynamics – CO2-EOR at the Forefront of Enhanced Oil Recovery Evolution". This chapter explores the dynamic evolution of enhanced oil recovery methods, showcasing the shift from thermal to CO2-EOR and the influence of reservoir lithology on method selection. It underscores the dual benefits of CO2-EOR in capturing significant CO2 while enhancing oil production.

Chapter 2, "Core Flooding Application in EOR WAG Implementation: The Impact of Asphaltene Precipitation and Deposition", investigates the impact of asphaltene precipitation and deposition on three-phase relative permeability data. The study provides a detailed examination through dynamic displacement experiments, shedding light on crucial insights for optimizing enhanced oil recovery strategies.

The exploration continues with Chapter 3, "Progress of CO_2 EOR and Storage Technology", offering a comprehensive overview of the promising CO_2 flooding technology. The chapter details the evolution of laboratory experiments, numerical simulation methods, and optimization strategies, emphasizing the multi-objective approach for both enhanced oil recovery and large-scale CO_2 storage.

Chapter 4, "In-Situ Synthesis of Nanoparticles for Enhanced Oil Recovery (EOR) Operations: Current Status and Future Prospects", offers insights into the in situ synthesis method and its advantages over ex-situ methods. The study examines the impact of in situ synthesized nanoparticles on oil production and crude oil upgrading, highlighting their potential to enhance the oil recovery factor.

Chapter 5, "Advancement in Hydraulic Fracturing for Improved Oil Recovery", provides a comprehensive overview of advancements in hydraulic fracturing, emphasizing innovations in fracturing fluids, proppants, perforation design, and diagnostic technologies. The chapter underscores the multifaceted nature of hydraulic fracturing advancements and their significance in the oil industry.

Chapter 6, "In-Situ Foam Generation: A Superior Method for Enhanced Oil Recovery in Unconventional Fractured Reservoirs", addresses the unique challenges posed by unconventional reservoirs. The study focuses on in situ foam generation under reservoir conditions, presenting an innovative approach to enhance oil displacement in heterogeneous systems.

Chapter 7, "Advancements and Operational Insights in the Bakken Shale: An Integrated Analysis of Drilling, Completion, and Artificial Lift Practices", provides a comprehensive exploration of the Bakken Petroleum System, focusing on drilling, completion, and artificial lift practices. The chapter evaluates the evolving techniques that have significantly enhanced oil recovery in this prominent unconventional resource basin.

Finally, Chapter 8, "Toward Understanding the Effect of Electromagnetic Radiation on In Situ Heavy Oil Upgrading and Recovery: Background and Advancements" explores the promising field of electromagnetic heating as a thermal enhanced oil recovery method. The study reviews the basic theory, mechanisms, and recent advancements, highlighting its economic feasibility and reduced environmental challenges.

In conclusion, this book is a valuable resource for researchers, professionals, and students interested in the forefront of enhanced and improved oil recovery. Each chapter contributes to collective knowledge and provides a glimpse into the innovative strides that continue to shape the industry. I trust that readers will find this volume both informative and inspiring.

Mansoor Zoveidavianpoor
Malaysia Petroleum Resources Corporation (UTM-MPRC),
Institute for Oil and Gas,
Universiti Teknologi Malaysia,
Johor, Malaysia

Chapter 1

Introductory Chapter: Lithological Dynamics – CO2-EOR at the Forefront of Enhanced Oil Recovery Evolution

Mansoor Zoveidavianpoor

1. Introduction

The evolution of Enhanced Oil Recovery (EOR) methods reveals a shift from thermal to CO2-EOR, with reservoir lithology influencing method selection. In sandstone, thermal and chemical methods like High-Pressure Air Injection (HPAI) and polymer flooding thrive. Conversely, carbonate reservoirs favor polymer flooding, with modest chemical contributions. Gas flooding methods, including CO2, find application in both lithologies. CO2-EOR emerges as a dual-benefit solution, capturing significant CO2 while boosting oil production. Despite challenges, CO2-EOR is globally recognized as an attractive carbon storage method.

Over the decades, the evolution of EOR methods is evident in the data presented in **Figure 1**. From the early years with a focus on thermal techniques, such as steam injection, to the growing significance of CO2-EOR and diverse chemical and thermal methods, the industry has witnessed dynamic shifts in project distribution and technology adoption.

The lithological composition of a reservoir is a screening factor that restricts the suitability of certain EOR methods [2]. The information presented in **Figure 2** adopted from Manrique et al. [3], consisting of 1450 projects, demonstrates the implementation of thermal, gas, and chemical EOR techniques in both sandstone and carbonate reservoirs. It clearly emphasizes the primary emphasis on EOR applications in sandstone reservoirs. According to **Figure 2**, sandstone reservoirs are more amenable to EOR projects due to extensive testing and successful evaluations of various technologies. In contrast, carbonate reservoirs, with low porosity and natural fractures, pose challenges like fluid bypassing. Gas injection is the primary method in carbonates, while polymer flooding stands as the sole proven chemical technique. Thermal methods have minimal contribution, but High-Pressure Air Injection (HPAI) is gaining traction, particularly in light oil carbonate reservoirs. In this chapter, a concise exploration of the three primary EOR methods—thermal, gas, and chemical—applied in both sandstone and carbonate reservoirs is provided.

IntechOpen

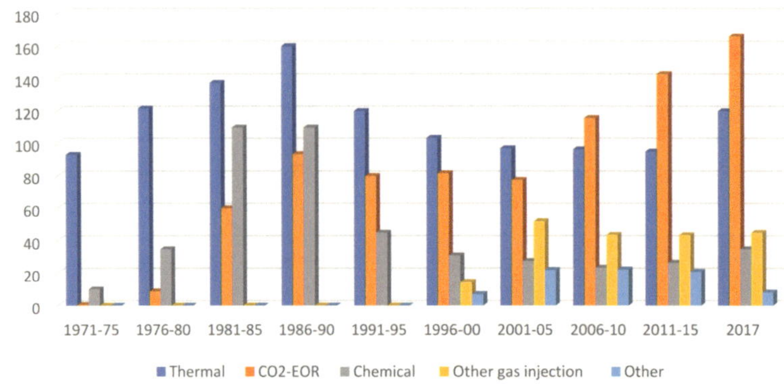

Figure 1.
Global distribution of enhanced oil recovery (EOR) projects, 1971–2017. Adapted from IEA [1], the data illustrates the evolution of EOR methods, showcasing trends and technology adoption over the decades.

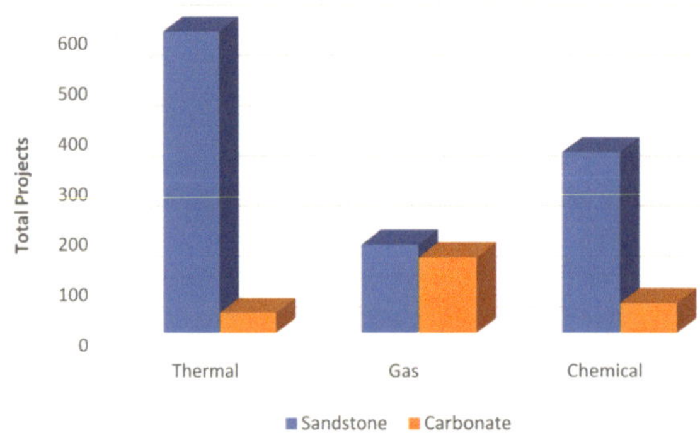

Figure 2.
EOR field projects categorized by reservoir lithology.

2. Thermal method: sandstone versus carbonate

In sandstone reservoirs, thermal EOR methods, such as cyclic steam injection, steam flooding, and Steam-Assisted Gravity Drainage (SAGD), are widely employed [4]. Optimization efforts involving solvents, gases, and additives for steam injection are ongoing, and SAGD stands out as a significant thermal EOR method [5].

In contrast, thermal methods, notably steam injection, exhibit limited application in carbonate formations. Nevertheless, according to Tang et al. [6], steam injection proves highly efficient for heavy oil reservoirs containing carbonate, and the authors outlined its potential recovery mechanism. Furthermore, as indicated by Xu et al. [7], thermal recovery through steam injection appears to be the preferred option for heavy oil reservoirs containing carbonate formations. Instances include steam drive projects in specific United States fields and small-scale testing in various international locations. HPAI has gained prominence in carbonate reservoirs, with successful projects in the United States, such as in Montana, North Dakota, and South Dakota, showcasing economic viability and controllable risks. Ongoing exploration of air injection processes, particularly HPAI, in carbonate reservoirs, suggests potential future developments influenced by the outcomes of the global recent projects [8].

3. Chemical methods: sandstone versus carbonate

In sandstone reservoirs, chemical EOR methods, notably polymer flooding, are well-established and globally implemented. Ongoing projects explore advanced technologies like Colloidal Dispersion Gels (CDGs) for improved sweep efficiency; however, as indicated by Wang and Seright [9], there is no credible evidence that CDGs can propagate deep into the porous rock of a reservoir. The development of Alkali-Surfactant-Polymer (ASP) technology, particularly in China's Daqing field [10], has reignited interest in chemical floods, evident in active ASP and surfactant-polymer (SP) projects globally. Despite oil price fluctuations, the industry maintains a strong focus on chemical flooding, with growing interest and projects in various countries like Argentina, Canada, India, and the United States [11].

In contrast, carbonate reservoirs primarily rely on polymer flooding as the proven chemical EOR technology, applied mostly during early waterflooding stages. However, carbonate reservoirs have modestly contributed to the success of polymer flooding [12]. While Alkali-Polymer (AP), SP, and ASP floods are exclusive to sandstone reservoirs, ongoing lab testing of ASP in carbonate formations suggests potential future applicability. Strategies based on chemicals, such as employing gels and foams for gas and water shutoff, are anticipated to enhance the efficiency of water, gas, or Water-Alternating-Gas (WAG) projects in carbonate reservoirs in the imminent future.

4. Gas methods: sandstone versus carbonate

In sandstone reservoirs, gas flooding methods are commonly used for light, condensate, and volatile oil, with nitrogen (N_2) injection proposed to vaporize light fractions. Ongoing projects, notably in Hawkins Field (Texas) and Elk Hills (California), utilize N_2 injection. Hydrocarbon gas injection, excluding pressure maintenance or double displacement, has marginal contributions, particularly on Alaska's North Slope, with potential differences in offshore sandstone reservoir dynamics. Non-hydrocarbon gases like N_2 and CO_2 provide alternatives, preserving reservoir pressure and enhancing oil recoveries. CO_2 flooding, widely used for medium and light oil in United States sandstone reservoirs, has successful global applications [13].

In carbonate reservoirs, gas flooding methods, including nitrogen and hydrocarbon gas injection, are widely employed for EOR. The majority of injection projects in carbonate reservoirs, accounting for 61%, utilize CO_2, while 36% involve hydrocarbon gas injection, and a mere 3% are dedicated to N_2 injection [7]. N_2 flooding, once effective, has waned in interest due to high costs, with HPAI emerging as a cost-effective alternative, especially in regions like Montana, North Dakota, and South Dakota. Hydrocarbon gas injection in onshore carbonate reservoirs has a marginal contribution to total oil recovery. Utilizing natural gas for pressure maintenance or in WAG and foam-assisted water-alternating gas (FAWAG) processes is considered practical, preserving reservoir energy and maximizing oil recovery [14]. CO_2-EOR in carbonate formations relies on natural CO_2 sources, especially in the United States Permian Basin, with expected growth unless more viable EOR strategies are developed [15]. Globally, CO_2-EOR is seen as an attractive CO_2 storage method, particularly from anthropogenic sources, despite challenges like limited storage capability and high costs. Additionally, acid gas injection, involving a mixture of H_2S and CO_2 [16], serves as an EOR strategy in carbonate formations, with ongoing or planned projects in various locations, including Zama field (Canada), Tengiz field (Kazakhstan), and Harweel (Oman).

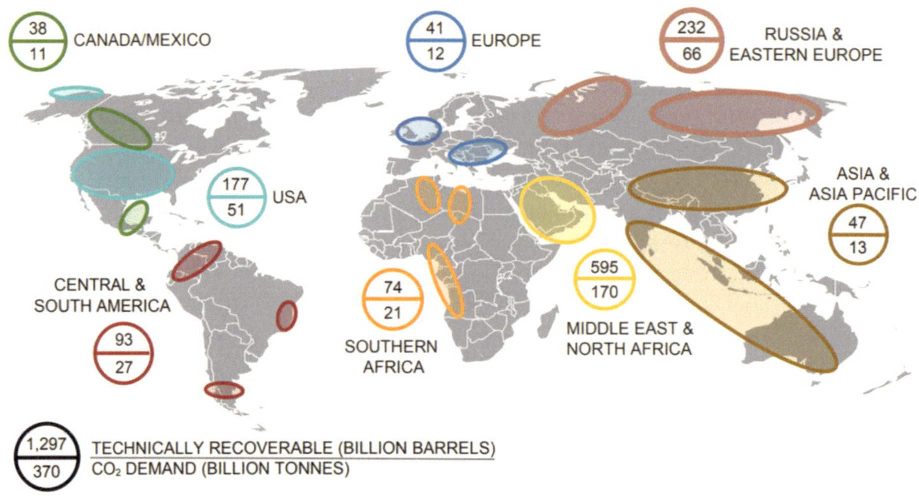

Source: IEA Greenhouse Gas R&D Programme, "CO_2 Storage in Depleted Oilfields: Global Application Criteria for Carbon Dioxide Enhanced Oil Recovery," December 2009.

Figure 3.
Global potential for EOR using CO2 in conventional fields [17].

CO2-EOR is a method capable of capturing significant amounts of CO2 while boosting crude oil production from established oil fields. This process yields oil with a potentially lower carbon footprint compared to conventional methods. CO2-EOR, employed for over four decades, utilizes a closed-loop injection and recycle system to trap CO2. Injected into reservoirs, CO2 serves as a solvent, expanding oil volume, reducing viscosity, and facilitating oil movement from injection to production wells. This established technique maximizes hydrocarbon recovery in new fields and prolongs the lifespan of mature oil fields. United States, Middle East, North Africa, Eastern Europe, and Russia contain most of the potential for CO2-EOR and its associated storage volume (see **Figure 3**).

In summary, the lithology of reservoirs significantly influences the selection of Enhanced Oil Recovery (EOR) methods. Sandstone reservoirs favor thermal and chemical approaches, with ongoing advancements such as HPAI. Chemical EOR, particularly polymer flooding, is widespread globally in sandstone. Carbonate reservoirs primarily rely on polymer flooding, with limited chemical contributions. Gas flooding methods, including N2 and hydrocarbon gas injection, are common in both sandstone and carbonate reservoirs. HPAI emerges as a cost-effective alternative in carbonate formations. CO2 flooding, successful in United States sandstone, is applied globally, including carbonate formations. Despite challenges, CO2-EOR is considered an attractive CO2 storage method. The evolving EOR landscape reflects ongoing projects and technological advancements in different lithologies.

Author details

Mansoor Zoveidavianpoor[1,2]

1 Faculty of Chemical and Energy Engineering, Department of Petroleum Engineering, Universiti Teknologi Malaysia, Johor, Malaysia

2 Malaysia Petroleum Resources Corporation (UTM-MPRC) Institute for Oil and Gas, Universiti Teknologi Malaysia, Johor, Malaysia

*Address all correspondence to: mansoor353@yahoo.com; mansoor@utm.my

References

[1] IEA. Number of EOR Projects in Operation Globally, 1971-2017. Paris: IEA; 2018. Available from: https://www.iea.org/data-and-statistics/charts/number-of-eor-projects-in-operation-globally-1971-2017 (accessed December 11, 2023); Licence: CC BY 4.0

[2] Bakhshi P, Kharrat R, Zallaghi M. The effect of lithology on EOR and CO2 geo-storage potentials of carbonated water flooding. In: EAGE 2020 Annual Conference & Exhibition Online. Vol. 2020(1). European Association of Geoscientists & Engineers; 2020. pp. 1-5

[3] Manrique E, Thomas C, Ravikiran R, Izadi M, Lantz M, Romero J, et al. EOR: Current status and opportunities. In: SPE Improved Oil Recovery Conference? Tulsa, Oklahoma, USA: SPE; 2010. p. SPE-130113

[4] Mokheimer EM, Hamdy M, Abubakar Z, Shakeel MR, Habib MA, Mahmoud M. A comprehensive review of thermal enhanced oil recovery: Techniques evaluation. Journal of Energy Resources Technology. 2019;**141**(3):030801

[5] Vishnumolakala N, Zhang J, Ismail NB. A comprehensive review of enhanced oil recovery projects in Canada and recommendations for planning successful future EOR projects. In: SPE Canada Heavy Oil Conference. SPE-199951-MS, Virtual: OnePetro; 2020

[6] Tang G, Inouye A, Lowry D, Lee V. Recovery mechanism of steam injection in heavy oil carbonate reservoir. In: SPE Western North American Region Meeting. Anchorage, Alaska, USA: Society of Petroleum Engineers; 2011

[7] Xu ZX, Li SY, Li BF, Chen DQ, Liu ZY, Li ZM. A review of development methods and EOR technologies for carbonate reservoirs. Petroleum Science. 2020;**17**:990-1013

[8] Khakimova L, Askarova A, Popov E, Moore RG, Solovyev A, Simakov Y, et al. High-pressure air injection laboratory-scale numerical models of oxidation experiments for Kirsanovskoye oil field. Journal of Petroleum Science and Engineering. 2020;**188**:106796

[9] Wang D, Seright RS. Examination of literature on colloidal dispersion gels for oil recovery. Petroleum Science. 2021;**18**(4):1097-1114

[10] Guo H, Li Y, Kong D, Ma R, Li B, Wang F. Lessons learned from alkali/surfactant/polymer-flooding field tests in China. SPE Reservoir Evaluation & Engineering. 2019;**22**(01):78-99

[11] Pothula GK, Vij RK, Bera A. An overview of chemical enhanced oil recovery and its status in India. Petroleum Science. 2023;**20**(4)

[12] Diab WN, Al-Shalabi EW. Recent developments in polymer flooding for carbonate reservoirs under harsh conditions. In: Offshore Technology Conference Brasil. Brasil: OTC; 2019. p. D031S033R008

[13] Ettehadtavakkol A, Lake LW, Bryant SL. CO2-EOR and storage design optimization. International Journal of Greenhouse Gas Control. 2014;**25**:79-92

[14] Gugl R, Kharrat R, Shariat A, Ott H. Evaluation of gas-based EOR methods in gas-invaded zones of fractured carbonate reservoir. Energies. 2022;**15**(13):4921

[15] Gardiner J, Thomas RB, Phan TT, Stuckman M, Wang J, Small M, et al.

Utilization of produced water baseline as a groundwater monitoring tool at a CO2-EOR site in the Permian Basin, Texas, USA. Applied Geochemistry. 2020;**121**:104688

[16] Ofori K, Barifcani A, Phan CM. Effect of higher H2S concentration over CO2 in acid gas mixtures during geosequestration. Discover Chemical Engineering. 2023;**3**(1):18

[17] IEA Greenhouse Gas R&D Programme (IEA GHG). CO2 Storage in Depleted Oilfields: Global Application Criteria for Carbon Dioxide Enhanced Oil Recovery; 2009

Chapter 2

Core Flooding Application in EOR WAG Implementation: The Impact of Asphaltene Precipitation and Deposition

Ahmad Khanifar and Mansoor Zoveidavianpoor

Abstract

Pressure depletion, temperature changes, and injection of CO_2 or solvents into reservoirs can induce asphaltene precipitation and deposition in porous media. The dynamic displacement efficiency of a water alternating gas (WAG) process is controlled by relative permeability. Asphaltene deposition may alter the original characteristics of relative permeability curves. In this study, the effects of asphaltene deposition on the three-phase relative permeability using dynamic displacement experiments are investigated. A synthetic experimental approach is used to simulate the effect of in situ asphaltene deposition on the three-phase relative permeability of a water-wet system. Independent core flooding experiments are conducted on the different core-plug samples. The experimental results show that the asphaltene deposition changes the system from water-wet to mixed-wet. However, the results for the gas-oil system indicate it does not have a significant effect on gas-oil relative permeability. The oil relative permeability in a three-phase system shows different trajectories for oil iso-perm with different levels of asphaltene deposition until a certain gas saturation is achieved. Beyond this saturation point, all oil iso-perm trajectories merge, indicating no significant effect of asphaltene deposition. Understanding the impact of asphaltene precipitation and deposition in WAG application can provide valuable insights for optimizing enhanced oil recovery strategies.

Keywords: asphaltene, formation damage, relative permeability, enhanced oil recovery, core flooding, reservoir engineering

1. Introduction

Asphaltene deposition is a severe problem that some oil reservoirs may face during their production life. It may occur during natural depletion, displacement of reservoir oil by CO_2 or hydrocarbon gas, or during WAG application. There are several studies in the literature that have addressed asphaltene problems during the primary recovery or CO_2 injection as a secondary recovery stage [1–3]. Despite all research and studies conducted in the past, the definition of asphaltene itself is not yet very well understood, and it has been defined based on its solution properties. Asphaltene is arbitrarily defined as a soluble class of petroleum that is insoluble in light alkanes such as n-heptane or n-pentane but soluble in toluene or dichloromethane [4].

An evaluation of the asphaltene stability is the first step toward predicting and avoiding any asphaltene issues at reservoir conditions. The asphaltene equilibrium conditions can be disrupted due to pressure depletion, temperature change, change in crude oil composition, and the addition of miscible gases and liquids to the oil as applied in various enhanced oil recovery (EOR) techniques [5]. The effect of the composition and pressure change on asphaltene precipitation is generally believed to be higher than the temperature [6]. The onset point of asphaltene is the point at which asphaltene loses its stability from thermodynamic equilibrium in solution and forms a separate and visible phase that starts the point of precipitation step [7]. During this step, asphaltenes that have tendencies to aggregate may reach the flocculation and then deposition steps. Indeed, after asphaltene precipitates from the oil, they may flocculate to form much larger-sized molecules; however, they are still suspended in the solution. The flocculated asphaltenes which can be suspended with oil flow may be deposited on the rock surface because they become so large and cannot be carried by the liquid [8]. Therefore, asphaltene deposition means the settling of asphaltenes' flocculated particles onto the rock surfaces. The flocculated asphaltenes can be adsorbed onto the rock surface by adsorption or may be trapped within the porous media because of their size, thereby blocking the pore throats of the formation by plugging. Moreover, the deposited flocculated asphaltenes can be flushed away by oil due to the shearing effect if the local oil velocity is high because of entrainment [9].

1.1 The impact of asphaltene on reservoir performance

Heavy organic components such as asphaltenes, resins, and waxes exist in crude oils in various quantities and forms. Such compounds could separate out of the crude oil solution because of various mechanisms and deposits. The reasons for the asphaltene deposition can be many factors including variations in temperature, pressure, pH, composition, flow regime, wall effect, and electrokinetic phenomena [10]. Many papers have addressed asphaltene problems during primary recovery or CO_2 injection as a secondary recovery stage [1–3, 9, 11].

Formation damage because of asphaltene deposition in the oil industry is an issue for many fields that causes a reduction in production and shutting off of some of the wells and a severe detrimental effect on the economics of oil recovery. Once the asphaltene deposition occurs, it may cause severe permeability and porosity reduction and wettability alteration, changing relative permeability in the reservoir and, in severe cases, plugging the wellbore and surface facilities [12–14]. The approach taken by most operators is a remedial solution rather than preventive one. The remedial measures such as chemical treatment and workover operations are disruptive and expensive [1]. Thus, the probability of asphaltene precipitation and deposition occurring during any EOR techniques, its effects on reservoir performance, and preventive measures should be anticipated at the earliest stages of each project. This anticipation can be reached through a better understanding of the mechanisms up front that initiate such problems [15].

1.2 Analyzing asphaltene effects in EOR WAG: numerical and experimental approaches

EOR processes can modify the flow and phase behavior of reservoir fluids and rock properties. These modifications could lead to asphaltene precipitation and deposition and cause formation damage problems [16, 17]. Asphaltene-deposited particles caused by impairing the permeability by plugging the pore throat and altering wettability by adsorbing on the rock surface may lead to formation damage [1]. Deposition of solid asphaltenes causes porosity and absolute permeability reduction. This can also result in the alteration of rock wettability from water-wet to mixed- or oil-wet and the plugging of the wellbore and piping in production facilities [16, 18]. Asphaltene deposition may induce favorable changes in relative permeability and end-point saturations; hence, it can affect the displacement efficiency [19]. The main mechanisms behind this alteration are still a research topic. However, it has been reported that some of its effects can be captured by wettability change and relative permeability shift from a water-wet to a mixed- or oil-wet system. Few technical published papers related to various EOR/IOR case studies including CO_2 gas injection, immiscible water alternating gas (IWAG), gravity-assisted simultaneous water alternating gas (GASWAG), chemical EOR (CEOR), chemical water alternating gas (CWAG), and recent carbon capture and storage (CCS) related papers from the authors of this paper are given for reference [1, 20–24]. To investigate the effect of asphaltene deposition on relative permeability during WAG application, dynamic experiments were designed and conducted in the core flooding system under reservoir conditions. Because of the experimental difficulty of three-phase relative permeability measurements, first, the effect of asphaltene on relative permeability is investigated in the water-oil system as well as the gas-oil system separately, and then they are combined obtaining a three-phase relative permeability system [7].

1.3 Pioneering researchers and outcomes in EOR WAG core flooding studies

Recent studies have shown that formation damage prevention, assessment, control, and removal are among the important issues dealing with the oil and gas production from petroleum reservoirs [25]. Any EOR process can modify the flow and phase behavior of the reservoir fluid and rock properties, and these modifications could lead to asphaltene precipitation and deposition and cause formation damage problems [16]. In cases of deposition, the damage area is normally restricted to the wellbore zone; but in the case of precipitation, damage extends over large distances from the wellbore.

Asphaltene precipitated particles by impairing the permeability through plugging pore throat and altering wettability through adsorbing on negatively charged mineral sites may lead to formation damage [1]. The deposition of solid asphaltenes causes a reduction of the pore space available for fluids, that is, porosity reduction, absolute permeability reduction, alteration of rock wettability from water-wet to mixed-wet (or oil-wet), and plugging of the wellbore and piping in production facilities [16]. The deposition may induce favorable changes in relative permeability, end-point saturations, and effects on the displacement efficiency [19]. For instance, wettability controls the flow and distribution of immiscible fluids in an oil reservoir, which plays a key role in any oil recovery process. One way that oil components are thought to alter wettability is by coating pore surfaces with precipitated asphaltenes. The deposited asphaltene may cause the alteration of rock wettability from water-wet to oil-wet. Wettability has been shown to affect relative permeability, irreducible water saturation, residual oil saturation, capillary pressures, dispersion, and electrical properties. The alteration of relative permeability and end-point saturations has the strongest influence on displacement processes.

It should be emphasized that alteration of wettability from water-wet to oil-wet is not necessary to cause formation damage for all reservoirs. This wettability change may improve displacement performance and efficiency and may be favorable for oil recovery depending on the nature of wettability [26]. In fact, the relation between recovery and wettability is very complex, and it is still a controversial subject. Kamath et al. [27, 28] studied the effect of asphaltene deposition on permeability, pressure drop, and displacement performance of oil by water using one core of consolidated Berea sandstone and two unconsolidated sand packs. The results showed that asphaltene deposition caused permeability reduction but also caused improvement of oil displacement by water because of the improvement in oil relative permeability. In conclusion, wettability alteration is itself a complex topic and is a subject of active research. In this study, the experimental results emphasize that the wettability alteration from water-wet to more oil-wet or mix-wet because of asphaltene deposition occurred.

1.4 Enhancing reservoir performance amid asphaltene challenges: insights from core flooding predictions

Relative permeability is a complicated function of fluids and rock properties. It is believed to be affected by the following factors: pore geometry, wettability, fluid distribution, saturation, saturation history, flow rate, viscosity ratio, temperature, overburden pressure, interfacial tension, density, initial wetting phase, and immobile phase saturation [29–31]. All these factors are reviewed in the relative permeability of petroleum reservoirs book written by [29]. They pointed out that all factors that influence flow in systems containing two mobile phases can apply to three-phase systems as well. Furthermore, they mentioned that the wettability of reservoir rock has a major impact on relative permeability curves and subsequent reservoir performance. They emphasize that wettability is the main factor responsible for the microscopic fluid distribution in porous media, and it largely determines the amount of residual oil saturation and the ability of a particular phase to flow. Moreover, many researchers have reported the effects of hysteresis and spreading coefficient on three-phase relative permeability. The problem of hysteresis increases significantly when moving from a two-phase to a three-phase flow system. The three-phase hysteresis problem is significantly more advanced than that in the two-phase flow for two reasons. First, the number of saturation directions increases and, second, the definition of hysteresis becomes ambiguous. On a macroscopic scale, the number of process paths increases from a two-phase flow to a three-phase flow. In addition, the saturation path within the ternary diagram is not predefined in three-phase systems. For the two-phase case, the only unknown part of the saturation trajectory is its endpoints, as compared with the three-phase flow for which the whole saturation trajectory is initially unknown. On a microscopic scale displacement, sequences that can occur in three-phase systems are not seen in two-phase systems. These include double displacement mechanisms and spreading behavior of the intermediate wetting phase. Depending on the equilibrium spreading coefficient, one or two phases can be distributed as films in the porous medium.

2. Methodology

In this section, we describe the core flooding set ups, experiment procedure, and materials used to conduct the experiments. The procedure of these core flooding experiments in water-oil system and gas-oil system are presented in **Figures 1** and **2**, respectively. For each system, first, the relative permeability was measured without asphaltene deposition as a baseline, and then the relative permeability under different asphaltene percentages was measured using a synthetic approach. The preparation of a proper oil sample is a crucial step in conducting any asphaltene study. In this study, because of the difficulty in the preparation of the down-hole sample, the surface oil sample is used. For the investigation of the asphaltene effect on relative permeability by this type of sampling, a synthetic dynamic procedure is used to simulate asphaltene deposition and study asphaltene problems associated with it. In this procedure, n-heptane solvent as an asphaltene precipitating agent is used to create in situ asphaltene precipitation in porous media during core flooding experiments.

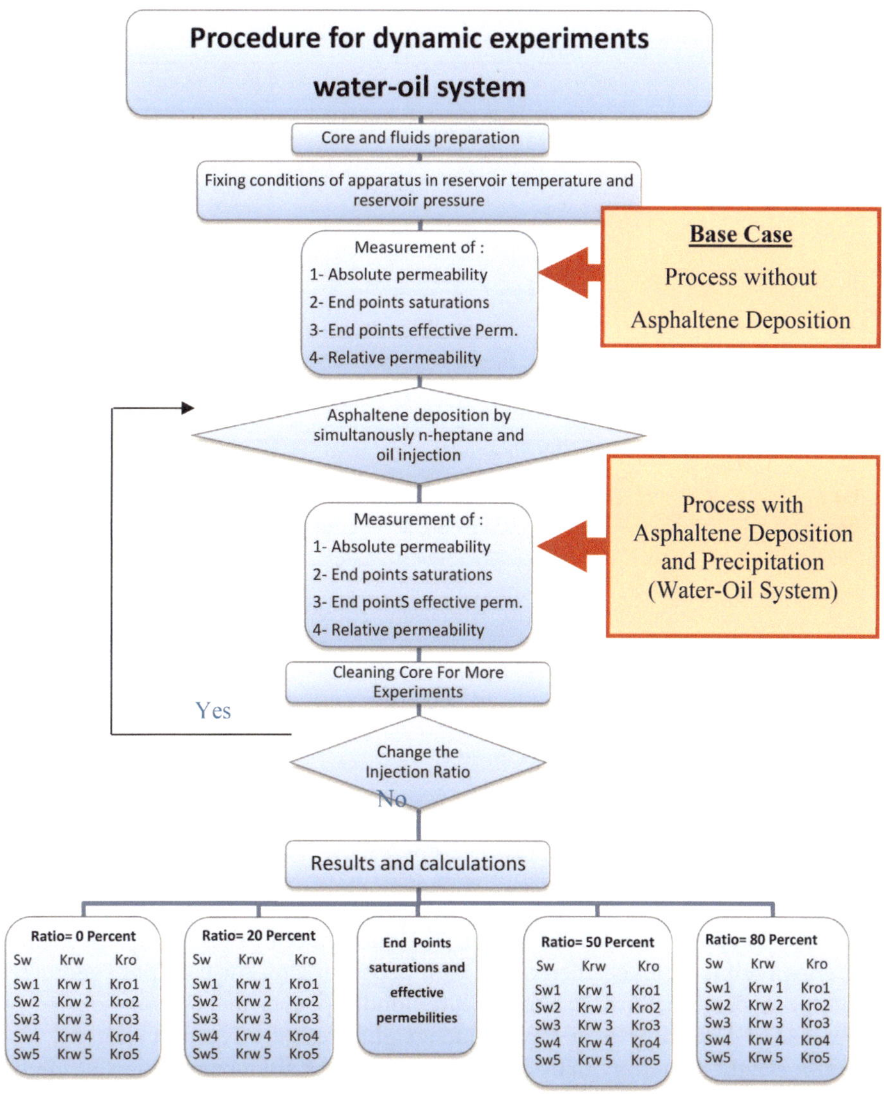

Figure 1.
General flowchart of core flooding experiments, water–oil system.

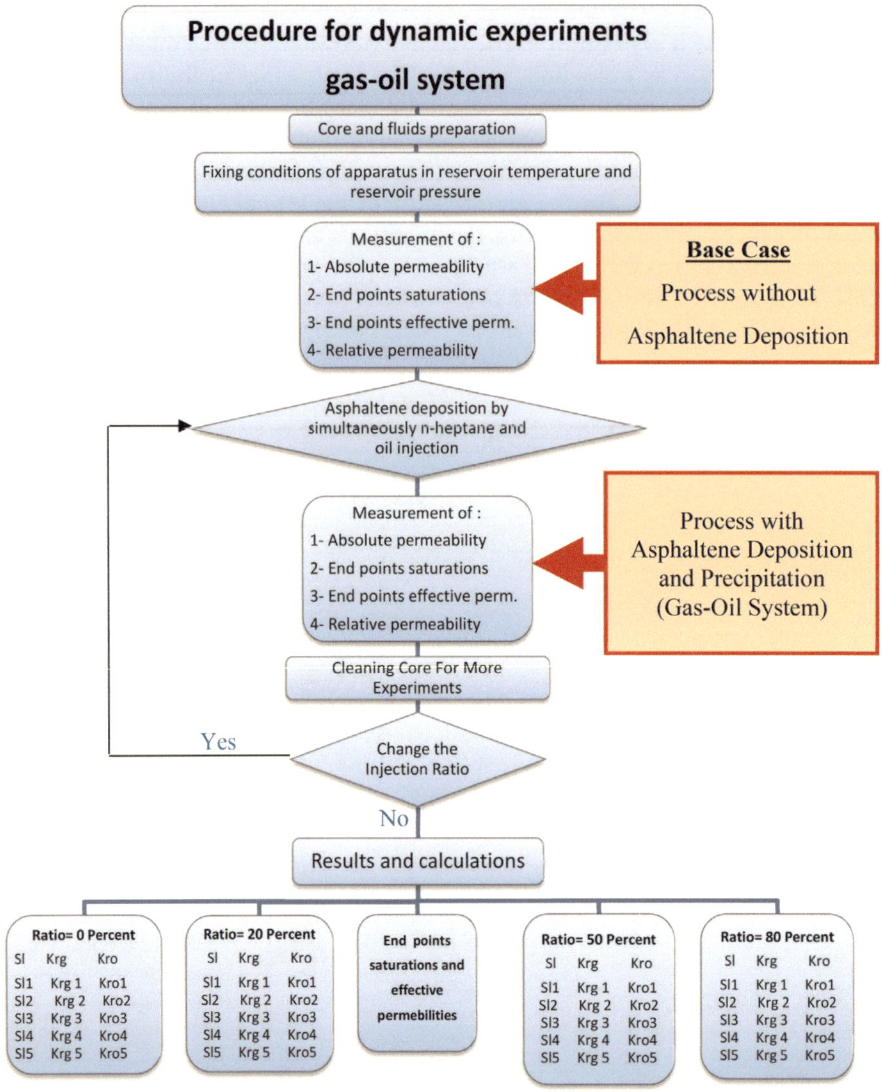

Figure 2.
General flowchart of core flooding experiments, gas-oil system.

As reported in the literature, n-alkanes are common solvents to precipitate asphaltene from dead oil [4, 7].

In this synthetic experimental approach, n-heptane and oil are injected simultaneously through different injection ports into a core sample. These fluids can be completely miscible inside porous media. The miscibility of these fluids may lead to asphaltene precipitation and then deposition depending on the core pore geometry and core pore-size distribution. During core flooding experiments, some oil samples are collected to determine their asphaltene weight presents. The reduction in asphaltene weight presents can be an indication of asphaltene deposition inside the porous media. On the other hand, different amounts of n-heptane to oil ratio injection could lead to the creation of different amounts of asphaltene deposition. Therefore, the different ratios of simultaneously n-heptane to oil injection are chosen to obtain different degree of asphaltene deposition inside the core. This synthetic experimental setup and procedure are explained in detail for the water-oil system and gas-oil system, separately.

2.1 Core flooding experiment setup and materials

The schematic of the experimental setup used in this study to investigate the effect of asphaltene deposition on relative permeability in water-oil and gas-oil systems is presented in **Figure 3**. The experimental setup from Sanchez Technologies consists of two injection pumps, three fluid stainless steel accumulators, a radial core holder, a three-phase separator and collecting pumps, an automatic fraction collector, pressure transducers, a back pressure regulator, and electronic valves. The system is equipped with a powerful and user-friendly data acquisition system. A radial core holder manufactured by Sanchez Technologies can accommodate cores with variable lengths up to 12 in and a constant diameter of 1.5 in. The core holder can be placed horizontally, vertically, or tilted depending on the experiment objective. In this study, it is placed horizontally during all experiments, and injection is carried out from the left- to right-hand side. There are two pressure taps in the inlet and outlet of the core and four temperature taps placed at the inlet, across two sections of the core and at the outlet. In addition, the core holder has three inlet ports: two for the liquid and one for the gas injection and one outlet port for fluid production. The inlet and outlet ports of the core holder are connected to pressure and temperature transducers that can control the pressure and temperature around the core sample. The inlet ports are also connected to the three accumulators and then to injection pumps.

A rubber sleeve is placed inside the core holder to apply confining pressure around the core sample. The confining pressure is applied around the core sleeve to prevent any fluid leakage or bypass around the core sample; during the displacement experiments, the confining pressure on the core is provided by injecting distilled water in the annular space between the core holder and rubber sleeve. The three accumulators are used to store oil, chemicals, and water for delivery under high pressure up to

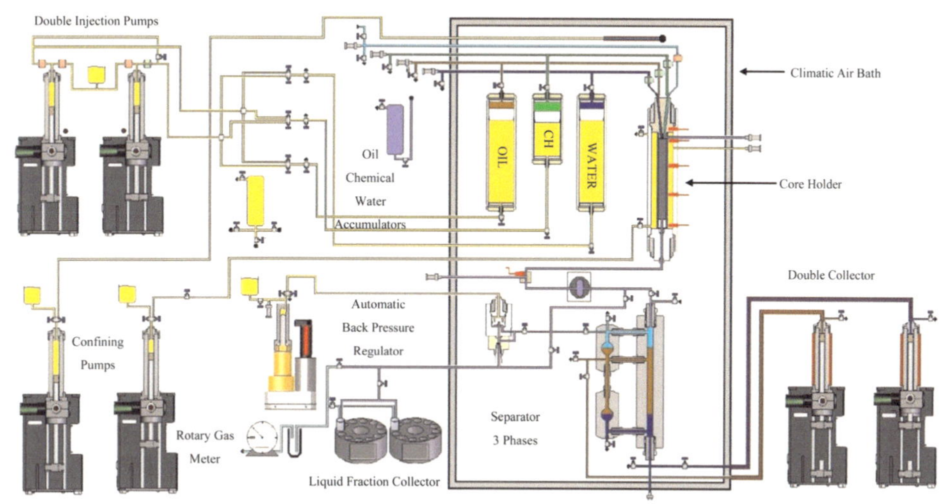

Figure 3.
Schematic of experimental setup used for displacement experiments.

15,000 psi. The oil and water columns are from stainless steel that can hold a volume of 3000 cc. The chemical accumulator can hold a volume of 1000 cc from Hastelloy materials that are resistant to corrosion. Three accumulators can work in a range of temperatures from ambient to 200°C. The outlet port can connect to an automatic fractional collector or three-phase separator through a back pressure regulator for the collection of produced fluid from the core sample. A back pressure regulator (BPR) can maintain a constant pressure in the system by blocking flow until the system pressure reaches the set pressure. The automatic fraction collector is used to collect the different fractions at the output of the differential valve continuously. It is composed of two heated trays, and each tray can accommodate three different sizes of collecting tubes with a capacity of 20, 15, and 10 cc. Each tray can hold twelve collecting tubes or collection liquid fractions. The temperature of the fraction collector can be increased from ambient to 80°C. It is worthwhile to mention that in a single experiment, it is not possible to use both separator and fraction collector simultaneously.

A three-phase separator is used to measure produced oil, gas, and water from core flooding experiments. The separator works based on an infrared detection system that varies with the physical properties of the phases. An infrared phase detection system includes a light source for emitting into a fluid. A detector can detect the attenuation of the infrared wavelength band as the infrared radiation passes through a fluid. Different fluids have a different but specific output signal that is used to determine the phase type. There are two infrared detection systems to determine the water-oil contact and gas-oil contact. At the beginning of experiments, both contacts should exactly be adjusted in front of the infrared system, as shown in **Figure 4**. Any changes in fluid level can be detected by signal changes. There are two collecting pumps and a wet rotary gas meter that can collect and measure the produced liquids and gases. A rotary gas meter that works upon the principle of positive

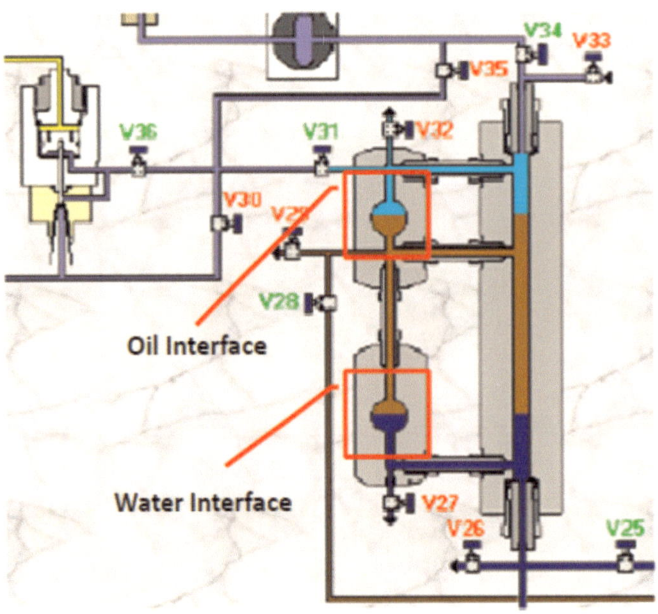

Figure 4.
Schematic of three-phase separator in core flooding system.

displacement measures the cumulative volume of gas released. The sample gas stream rotates a measuring drum within a packing fluid, usually water or low viscous "white" (clear) oil. A needle dial and a counting mechanism, coupled to the rotating drum, record the volume of gas flow as it sequentially fills and empties from the drum's rigid, fixed volume-measuring chambers. The complete core flooding system is placed inside the digital oven system; therefore, the system temperature can be controlled and kept at a certain temperature. For this study, the temperature of all core flooding experiments is kept around 60°C. In addition, the back pressure regulator is always used to keep the core flooding system pressure around 1500 psi.

The following sections (i.e., 2.2. and 2.3.) describe the core flooding experiments conducted for oil-water and gas-water systems.

2.2 Core flooding experiment in water-oil system

The unsteady-state or dynamic displacement method is the most frequently applied method in reservoir analysis of strong wetting preference and with homogeneous samples. Therefore, in this study, the unsteady-state method based on [32] is used for the determination of relative permeability data. In each of the following core sample, the subsequent experimental procedure is separately applied to investigate the effect of asphaltene precipitation and deposition on relative permeability and reservoir performance in the water-oil system.

- Determination of basic core properties: The bulk volume of the core is computed from the measurements of dimensions of the uniformly shaped core sample. After weighing the dry core sample, a brine of 10,000 ppm sodium chloride concentration is used for the saturation of the core sample. The saturation is done by using a vacuum pump and extracting out the initial fluid inside the sample and replacing the void pore spaces of the sample with prepared brine. The core sample is weighed again to determine the wet weight of the core sample. The value of pore volume is computed from the difference between the dry and the wet weight of the rock sample and brine density.

- Preparation and setting-up system: The saturated core sample is inserted into rubber sleeve and carefully tightened in the core holder to ensure direct contact between the core sample and core holder end pieces. The three accumulators are filled with crude oil sample, brine water, and n-heptane. The core flooding system temperature is increased to 60°C, and the overburden pressure is applied on the rubber sleeve that is always 500 psi over the injection pressure. The back pressure regulator is used to control the core flooding system pressure around 1500 psi.

- Absolute permeability, irreducible water, and initial oil saturation measurements: The brine is injected into the core at a constant flow rate of 1.0 cc/min until the pressure drop across the core is stabilized and a steady-state condition is attained. The absolute core permeability is calculated using Darcy's law and the stabilized pressure drop. Then, the oil is injected into the core at the same system injection pressure condition to displace brine water at a rate of 0.5 cc/min, until the pressure drop is again stabilized, and no more water can be displaced from the core. The total amount of brine production is recorded during the oil injection process. Again, the stabilized pressure drop is used to calculate effective oil permeability at irreducible water saturation by using Darcy's law. This is the original effective oil permeability without asphaltene effects at irreducible water saturation. The irreducible water saturation and initial oil saturation are then determined from the volumetric material balance.

- In situ asphaltene precipitation and deposition: As stated previously, the ratio of n-heptane to oil injection can control the amount of asphaltene precipitation and deposition. The three ratios 20, 50, and 80% are selected to create different asphaltene precipitation and deposition. The first core plug sample is used for the case without asphaltene precipitation with a 0% ratio n-heptane to oil injection. The three other core plug samples that are cut from the long core sample and with the mostly same properties are used for the ratios of 20, 50, and 80%, respectively. All steps from (a) to (f) are separately done for all core plug samples except for the first core plug sample, but Step (d) is excluded. The n-heptane is injected simultaneously along with the oil injection but in separate port injection with preselected ratio. During this process, the injection pressure and pressure drop are recorded and monitored. An increase in the injection pressure is an indication of asphaltene deposition. Some oil samples are collected for asphaltene weight percent determination during the simultaneous injection. This simultaneous injection is stopped after several hours and continued by only oil injection until the pressure drop once again stabilized and the core is saturated fully with oil. The material balance for asphaltene weight percent between the collected oil samples and injected oil is done to determine the amount of asphaltene deposited inside the core for this simultaneous ratio injection.

- Relative permeability measurements: The oil injection is stopped after the core is fully saturated with oil at irreducible water saturation. The brine injection has again started to displace oil out at a rate of 0.5 cc/min and at the same system injection pressure condition. Time, pore volume injection, pressure drop across the core, and oil and water production are measured continuously until the pressure drop is again stabilized and no more oil can be displaced from the core. Effective water permeability at irreducible oil saturation is again calculated based on the stabilized pressure drop and using Darcy's law. Irreducible oil saturation and final water saturation are then determined from the volumetric material balance, which is later explained in detail. The commercial core flow simulator [33], a black-oil simulation model, is used to determine the oil and water's relative permeability. This simulator performs history matching of the experimental data (e.g., differential pressure, oil production, and water production) obtained in the lab to estimate the relative permeability data.

- Cleaning: After measurements of relative permeability by dynamic displacement of oil by water, the core and core flooding system are flashed again with n-heptane, and the core is cleaned with toluene to extract the residual crude oil with asphaltene and water. Then, the core is dried.

2.3 Core flooding experiment in gas-oil system

The unsteady-state or dynamic displacement method is again used to investigate the effect of asphaltene on the relative permeability in the gas-oil system. In each of the following core sample, similar subsequent experimental procedures have been explained for the water-oil system except for step one. Indeed, Step (e) is separately applied to investigate the effect of asphaltene precipitation and deposition on relative permeability and reservoir performance in the gas-oil system.

During dynamic experiments in the gas-oil system, instead of using brine injection in Step (e), the gas injection should be used. Nitrogen is used for the gas injection process. For this purpose, after draining out all brine from the brine accumulator, the nitrogen is transferred carefully from the high-pressure cylinder into the accumulator at a pressure less than 1500 psi. After heating the nitrogen to the system temperature, indeed 60°C, the accumulator pressure reached 1500 psi exactly. Therefore, in these experiments, the gas injection pressure is adjusted around 1500 psi such as the oil injection pressure during the oil-water system. The gas is injected by setting a rate of 0.5 cc/min for the injection pump. The injection pump can inject water at a rate of 0.5 cc/min, and this injection can move the piston in the bottom of the accumulator and, as a result, the gas can be injected into the core sample. Because of gas compressibility, the pressure inside this accumulator should be constant and around 1500 psi until it can be assumed that this is the same as the gas injection rate. The gas rate during gas injection should represent the typical reservoir gas velocities and pressure drops of 1–5 psi/ft. For this purpose, after stopping oil injection and when the core is fully saturated with oil at irreducible water saturation, gas injection is started to displace oil out at a rate of 0.5 cc/min and at the same system injection pressure condition. Here, the time, the pore volume injection, the pressure drop across the core, and the oil and gas production are measured continuously until the pressure drop is again stabilized, and no more oil can be displaced from the core. Effective gas permeability at irreducible liquid saturation is again calculated based on the stabilized pressure drop and using Darcy's law. Irreducible liquid saturation and final gas saturation are then determined from volumetric material balance.

3. Result & discussion

3.1 Experimental results

The experimental results that are directly measured during core flooding experiments in water-oil and gas-oil system mostly include pressure drop across the core, oil production, and water production versus time. An example of these results using 20% ratio of n-heptane–crude oil injection in water-oil system is shown in **Figures 5–7** [34].

The end-point saturations during core flooding experiments for water, oil, and gas are computed based on the material balance calculation. The endpoints' effective and relative permeability for water, oil, and gas are computed based on the amount of pressure drop data during steady-state conditions, rock and fluid properties, and using Darcy's law. **Table 1** shows these results for the water-oil system [34].

The entire curves of relative permeability for water-oil and gas-oil systems are estimated during history matching processes of experimental data and using a one-

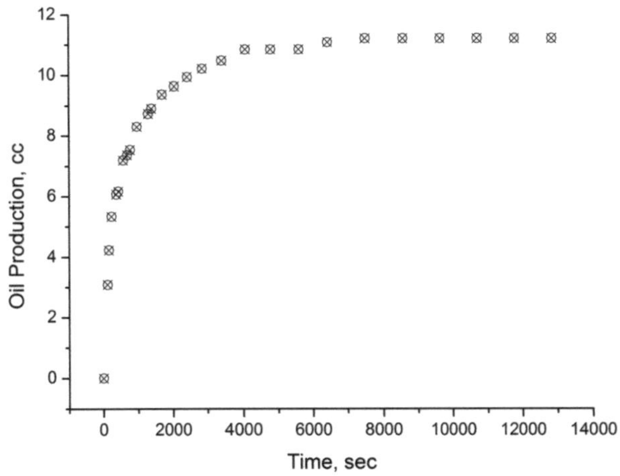

Figure 5.
Oil production (20% ratio of n-heptane–crude oil injection, water–oil system).

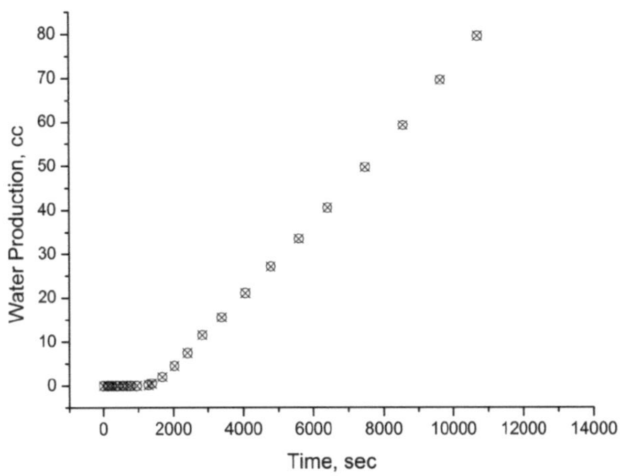

Figure 6.
Water production (20% ratio of n-heptane–crude oil injection, water–oil system).

dimensional, two-phase black-oil simulation model. Stone's II model is used to compute the three-phase relative permeability for the oil phase. The nonlinear multiregression analysis based on matching process results is used to develop the appropriate correlations for irreducible water saturation, residual oil saturation, water relative permeability, gas relative permeability, and oil relative permeability as a function of asphaltene deposition. Based on these observed experimental data, the cumulative pore volumes of oil production versus the cumulative pore volume of the gas injection at various ratios of n-heptane to crude oil injections are computed and can be shown in **Figure 8**. As shown, asphaltene deposition does not have significant effects on the oil production curve during gas injection. This may relate to the behavior of gas in sweeping up the oil. Indeed, in most systems, gas can play a non-wetting phase rule compared to oil and water. Whereas gas can enter the large pore spaces and can sweep up the oil better than water [34].

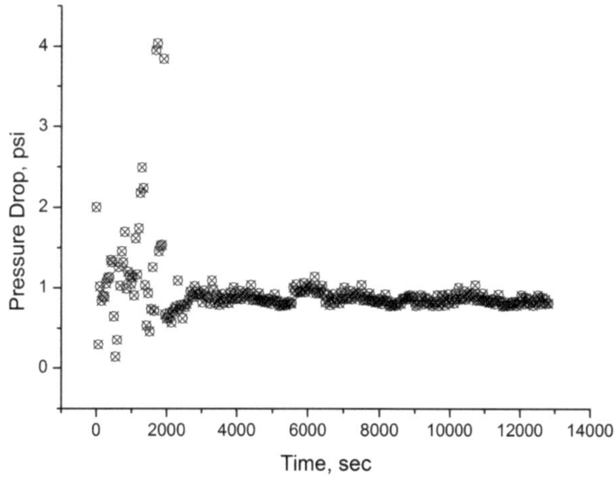

Figure 7.
Pressure drop (20% ratio of n-heptane–crude oil injection, water–oil system).

Ratio of n-heptane to oil injection, %	Length, cm	Diameter, cm	Oil viscosity, cp	Oil rate, cc/min	Pressure drop, psi	$k_{eo}(S_{wi})$, md	$k_{ro}(S_{wi})$, fraction
0	7.62	3.81	4.70	0.50	1.50	256.67	0.9589
20	7.65	3.81	4.70	0.50	2.00	192.67	0.7198
50	7.62	3.81	4.70	0.50	2.25	171.11	0.6393
80	7.60	3.81	4.70	0.50	2.51	153.39	0.5731

Table 1.
Effective and relative oil permeability (water–oil system).

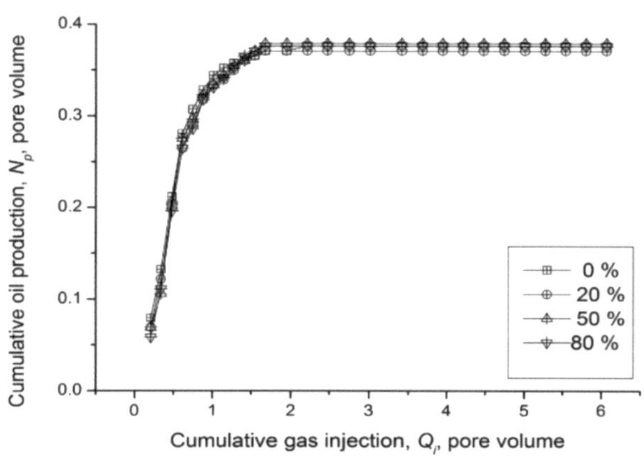

Figure 8.
Oil production versus gas injection (gas-oil system).

3.2 Modeling of relative permeability

A commercial core flow simulator is used to determine oil and gas relative permeability in the presence of irreducible water saturation [33]. The simulator performs history matching of the experimental data (e.g., differential pressure, oil production, water production) obtained in the lab to estimate the relative permeability data in the gas-oil system.

It is used as a one-dimensional, two-phase simulation model for analyzing special core analysis experimental data. This simulator is useful for the analysis of experimental data in the water-oil system as well as in gas-oil system; in addition, the two processes drainage and imbibition can be considered. It is worth noting that the simulator is equipped with well-known correlations of relative permeability such as Corey, LET, Burdine, Chierici, and Sigmund & McCaffery for estimation purposes [35, 36]. Experimental data versus time are introduced into software such as differential pressure across the core, oil and water production for the water-oil system and differential pressure across the core, oil production, and gas production for the gas-oil system obtained in the lab during the experimental performance. Then, one correlation such as the Corey correlation is chosen, and four endpoints of the relative permeability curves are introduced. Typically, an estimation step should preferably be initiated with one correlation and then proceed with other more correlations until an adequate history match of the experimental data is obtained. **Figure 9** shows a plot window with several plots after the matching process in this software.

Stone's II model is used to compute the oil relative permeability in a three-phase system based on two sets of two-phase relative permeability data. The relative permeability data in water-oil system and gas-oil system are obtained using the core flow simulator. The ternary diagram with fluid saturation points and iso-perm curves are commonly used to illustrate changes in the oil relative permeability values in the three-phase system. The computed values for the oil relative permeability for 20% n-heptane–crude oil ratios injection experiments based on Stone's II model are shown by oil iso-perm curves in **Figure 10**. The effect of asphaltene deposition on oil relative permeability is related to the amount of gas saturation. The oil's iso-perm in a three-phase system shows different trajectories with different levels of

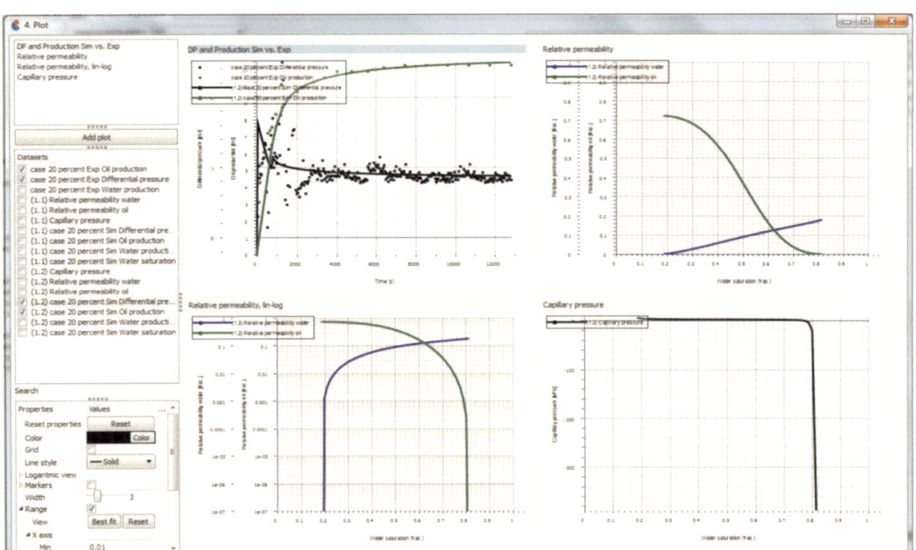

Figure 9.
Plot window with several plots after matching process.

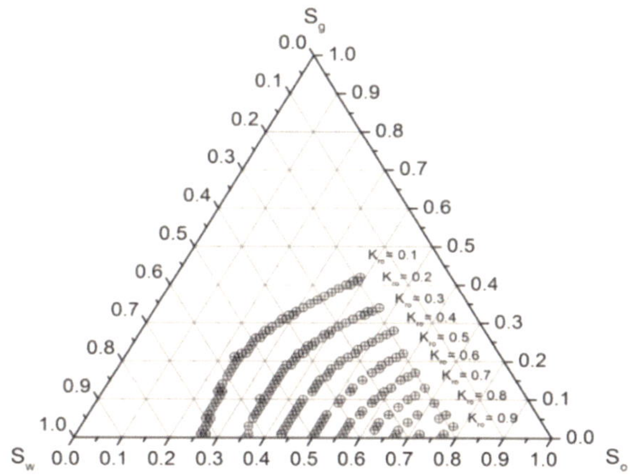

Figure 10.
Oil relative permeability at 20% ratio of n-heptane–crude oil injection.

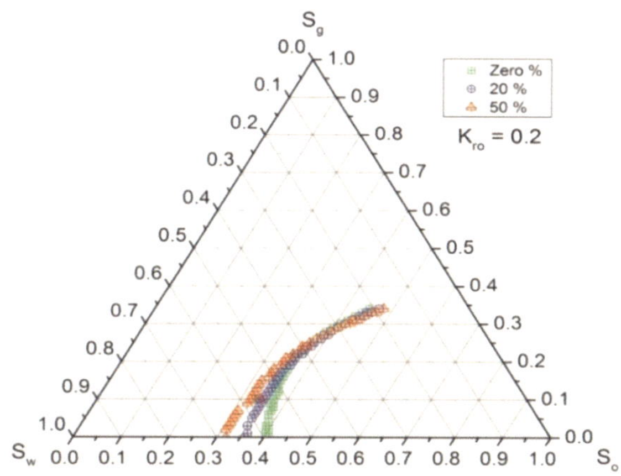

Figure 11.
Comparison of oil relative permeability of 0.2 for all cases.

asphaltene deposition until a certain gas saturation. For gas saturations above this saturation, all oil iso-perm trajectories merge, indicating no significant effect of asphaltene deposition. **Figure 11** illustrates a gas saturation level near 0.25 for an oil iso-perm of 0.2. Notably, two distinct behaviors are observed for oil iso-perm trajectories around this gas saturation. Above 0.25 gas saturation, all oil iso-perm trajectories merge, indicating no substantial impact from asphaltene deposition. However, below this gas saturation level, various oil iso-perm trajectories can occur, signifying a significant effect of asphaltene deposition.

3.3 Asphaltene modeling and numerical simulation

In this section, we model asphaltene deposition in porous media using numerical simulation with the experimental data obtained from the experiments described in the previous section. We use an eclipse commercial simulator.

3.3.1 Grid model & simulation inputs

Reservoir simulation has become a standard predictive tool in the oil industry. It can be used to obtain accurate performance predictions for a hydrocarbon reservoir under different operating conditions. There are several asphaltene models currently in use by simulators, but there is still no consensus about the characterization of asphaltene behavior. Basically, asphaltene modeling and simulation processes are decomposed into different stages in each simulator. Precipitation triggers a sequence of flocculation, deposition, and formation damage, including porosity and absolute permeability reduction, viscosity changes, and relative permeability alteration. A one-dimensional model with a grid dimension of $100 \times 1 \times 1$ is chosen. The uniform widths of each grid block in the x and y directions are 80 ft. with a uniform vertical grid block thickness of 20 ft. The porosity is 22.4% and the same for all grid blocks. The absolute permeability in the × direction is 260 md. The porosity and absolute permeability values are considered the same as those of the core sample properties. The water and CO_2 injector wells are located at Block 1 that is at the left edge of the reservoir, and the producer well is located at Block 100 that is at the right edge of the reservoir. **Figure 12** shows a schematic of the simulation model for this reservoir. The injection and production plan include 500-day natural depletion, 500-day water injection, and 2000-day cycle of WAG injection. The total recovery period is more than 8 years. The producer operates under a constant bottomhole pressure (BHP) of 500 psi. The water injection and gas injection wells commenced at a constant surface rate of 100 STB/day and 500 MSCF/day, respectively [34].

In the lack of using the live oil sample for this study, the fluid properties and the experimental asphaltene precipitation data required for building the fluid model are taken from [37]. The composition of this oil is given in **Table 2**. The oil contains 16.08% (weight) asphaltene in stock tank condition, a reported bubble point pressure of 2950 psi, and a stock tank oil API gravity of 19.0. Moreover, the experimental data of asphaltene precipitation have been reported for the oil sample at 212°F as a function of pressure and are shown in **Table 3** [34].

3.3.2 Asphaltene control parameters

One of the important steps for building a simulation model with asphaltene option is the introduction values for asphaltene control parameters. Typically, they should be adjusted based on core flooding experiment data. The sensitivity studies on the asphaltene control parameters by using the numerical study in two conventional simulators have been done [7]. They have shown that all asphaltene control parameters have affected the reservoir performance during natural depletion, water injection, and WAG application. Moreover, in the literature, there are only a few published core flooding experimental data related to asphaltene deposition and a few studies

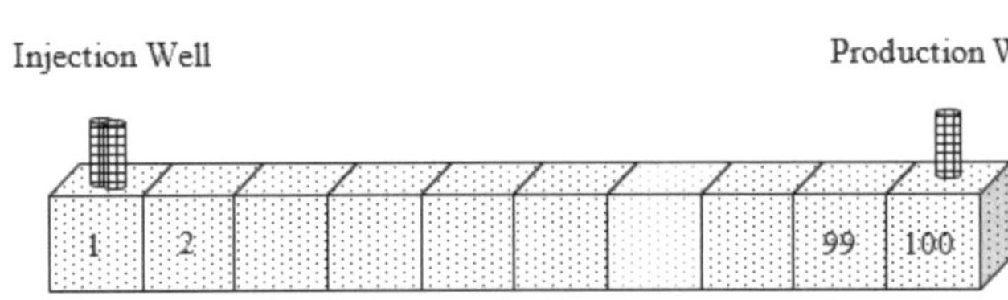

Figure 12.
Grid model for numerical simulation.

Component	Mole Fraction
Nitrogen	0.57
Carbon Dioxide	2.46
Methane	36.37
Ethane	3.47
Propane	4.05
i-Butane	0.59
n-Butane	1.34
i-Pentane	0.74
n-Pentane	0.83
Hexanes	1.62
Heptane plus	47.96
Total	100.00
C7+ molecular weight	329
C7+ specific gravity	0.9594
Live oil molecular weight	171.4
Stock tank oil API gravity	19.0
Asphaltene content in stock tank oil	16.8 wt%
Reservoir temperature	212°F
Saturation pressure	2950 psi

Table 2.
Experimental fluid properties.

Test pressure, psi	Precipitates live oil, wt%	Precipitates residual STO, wt%
1014.7	0.403	15.73
2014.7	1.037	14.98
3034.7	0.742	15.06
4014.7	0.402	14.86

Table 3.
Experimental asphaltene precipitation at 212°F.

presented procedures for adjusting the asphaltene control parameters [9, 14]. In this study, the asphaltene control parameters are adjusted based on the evaluation of values of absolute permeability reduction as a function of asphaltene deposition data from core flooding experiments and simulation. The equivalent values for asphaltene deposition during core flooding experiments are computed based on injection rate, total time of simultaneous injection, and average asphaltene weight present in collected oil samples. These values are shown in **Figure 13** by scatter points as denoted from the experiment. Typically, the absolute permeability can be correlated to porosity. Therefore, the reduction in absolute permeability because of asphaltene deposition can also be considered using a parameterized power law relationship given the

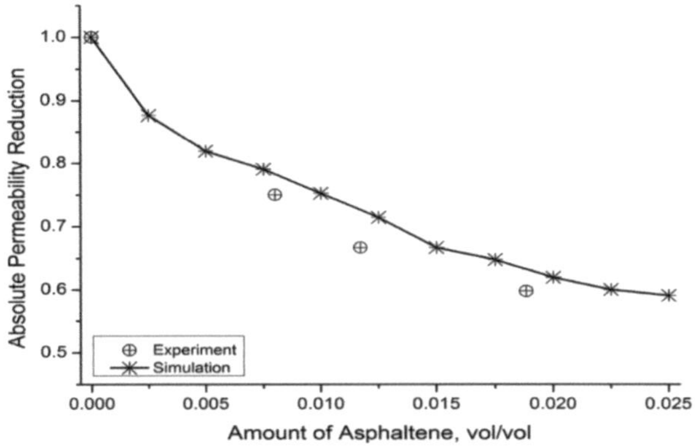

Figure 13.
Absolute permeability reduction matching between experiments and simulation.

ratio of the instantaneous permeability at time t with respect to the initial porosity and permeability and the volume fraction of asphaltene deposit [25, 34].

Therefore, the asphaltene control parameters have been adjusted during a history matching process of the experimental data with those predicted from the numerical simulation. During this attempt, the related values for asphaltene control parameters are changed until an acceptable match between experimental and simulation results are obtained. The comparison between the predicted values from simulation and experimental results of permeability reduction because of asphaltene deposition are shown in **Figure 13**. Moreover, the obtained values for the asphaltene control parameters by this attempt are shown in **Table 4** and used for further asphaltene simulations [34].

3.3.3 Relative permeability alteration

The relative permeability alteration during simulation can be considered by using the approach of weight factor *F* as a function of asphaltene deposit. To use this

Asphaltene Control Parameters	Value
Flocculation Rate Coefficient (day-1)	0.150
Dissociation Rate Coefficient (day-1)	0.001
Adsorption Coefficient (day-1)	0.100
Plugging Coefficient (ft-1)	0.100
Critical Deposition Fraction	0.0
Critical Flocs Concentration	0.0
Permeability Reduction Exponential Index	3
Entrainment Coefficient (ft-1)	0.0
Critical Velocity for Entrainment (ft/day)	2500
Constant of Generalized Einstein Model for Viscosity(Schlumberger, 2011)	2.5

Table 4.
Adjusted asphaltene control parameters.

approach, a table of weight factor F as a function of asphaltene deposition (α^*) and a set of oil-wet relative permeability data should be introduced into simulation file data. The simulator computes the amount of asphaltene deposition in each time step, and it uses Eqs. (1)–(4) to obtain the corresponding relative permeability data for the obtained amount of asphaltene deposition [34]:

$$S_{ora}(\alpha^*) = F \times S_{oro}(oil - wet) + (1 - F)(water - wet) \tag{1}$$

$$S_{wia}(\alpha^*) = F \times S_{wio}(oil - wet) + (1 - F) \times S_{wiw}(water - wet) \tag{2}$$

$$k_{rwa}(\alpha^*) = F \times k_{rwo}(oil - wet) + (1 - F) \times k_{rww}(water - wet) \tag{3}$$

$$k_{roa}(\alpha^*) = F \times k_{roo}(oil - wet) + (1 - F) \times k_{row}(water - wet) \tag{4}$$

In this study, to find suitable data for this approach based on the available experimental results, some assumptions must be considered. It is assumed that the relative permeability data obtained during 0% n-heptane-oil injection core flooding experiment are considered as water-wet relative permeability data. Therefore, the corresponding value for the weight factor F for this case is considered equal to zero ($F_1 = 0.0$). Furthermore, the relative permeability data obtained during 80% n-heptane-oil injection core flooding experiment are considered as oil-wet relative permeability data; consequently, the corresponding values for the weight factor F are considered to be equal to one ($F_4 = 1.0$).

As a result, the values of the weight factor F for the other two core flooding experiments (20 and 50% n-heptane-oil injection) that amount to their asphaltene deposition are expected between these two cases and should be between zero and one. By using the nonlinear, multiregression analysis for the values of the weight factor F as functions of the amount of asphaltene deposition, the following correlation can be provided to predict the other values for F.

$$F(\alpha^*) = \frac{\alpha^*}{A + B\alpha^* + C\sqrt{\alpha^*}} \tag{5}$$

where the coefficients in Eq. (5) are as follows: A = 0.00582525930049594, B = −0.0914083050003412, C = 0.0910209660724414, and α^* is the amount of asphaltene deposition [34].

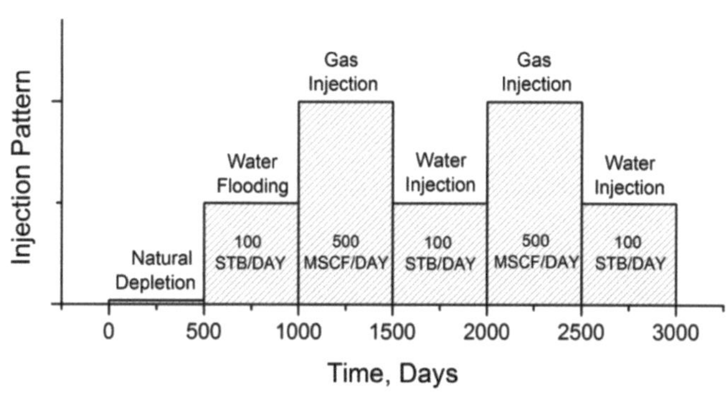

Figure 14.
Injection pattern during this study simulation.

3.3.4 Numerical simulation results

The injection pattern that has been conducted during this simulation is shown in **Figure 14**. The waterflooding started after 500 days of natural depletion. The two cycles of WAG implementation with 500 days of slugs as the EOR method are considered after the waterflooding process. In this simulation study, two different simulation cases are run to investigate the effect of asphaltene on reservoir performance during WAG implementation. The first case is without considering the asphaltene option, and the second case is with activating the asphaltene option [34].

Figure 15 shows the effect of the alteration of relative permeability data from a water-wet to more oil-wet system because of asphaltene deposition on the field oil efficiency based on the production well. As expected, and as can be seen in this figure, the field oil recovery factor for the asphaltene case is almost lower than the case without asphaltene. However, the ultimate oil recovery factor for the asphaltene case is higher than the case without asphaltene. It should be noted that this amount of the ultimate oil recovery factor is achieved by more than three pore volume injections. Furthermore, the core flooding results for oil recovery follow these simulation results. However, the question of how practical it is to inject fluid volumes of more than two pore volumes to achieve improvement in oil recovery in the presence of asphaltene deposition remains an important question to answer.

The field oil efficiency based on the production well, the field average pressure, the field oil production rate, and the well gas oil ratio (GOR) for these two cases are investigated (see **Figures 16–20**). It is shown asphaltene deposition in a porous medium has affected the reservoir performance significantly. For example, **Figure 16** shows that the field average pressure values for the asphaltene case are higher than the case without asphaltene. It can be noted that the asphaltene deposition by a reduction in porosity and absolute permeability can cause an increase in average oil reservoir pressure. Moreover, as can be seen, the most increases in the average reservoir pressure values are obtained during CO_2 gas injection slugs during WAG implementation periods. These can be related to an increase in the amount of asphaltene deposition because of CO_2 gas injection periods. **Figure 17** demonstrates that the maximum field

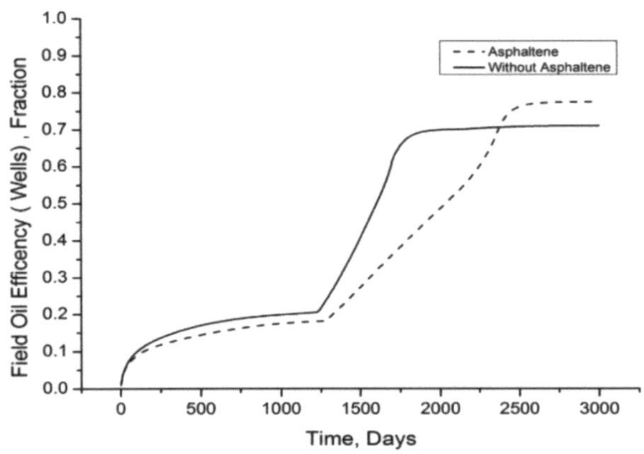

Figure 15.
Field oil efficiency factors.

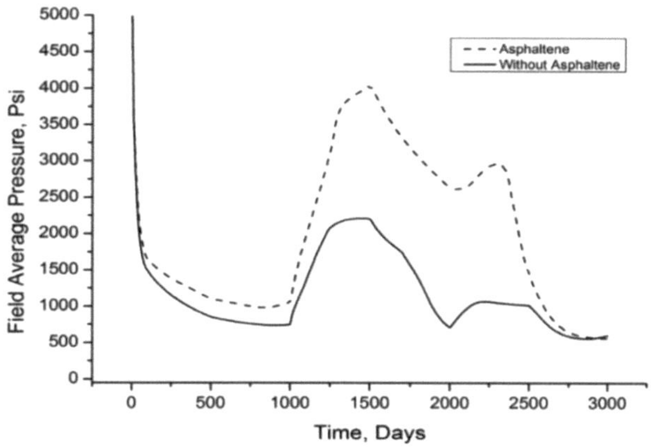

Figure 16.
Field average pressure.

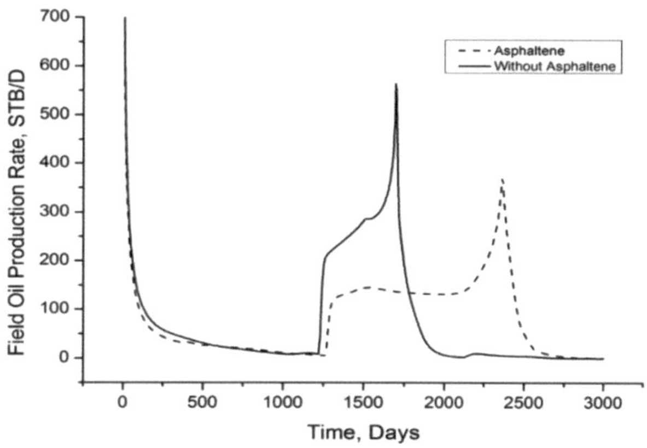

Figure 17.
Field oil production rate.

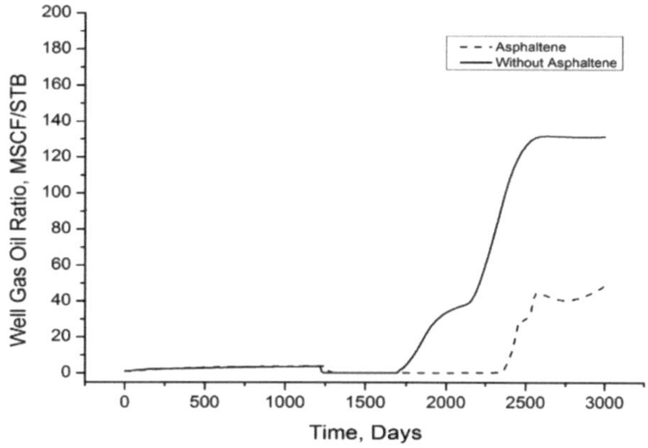

Figure 18.
Well gas oil ratio (GOR) for production well.

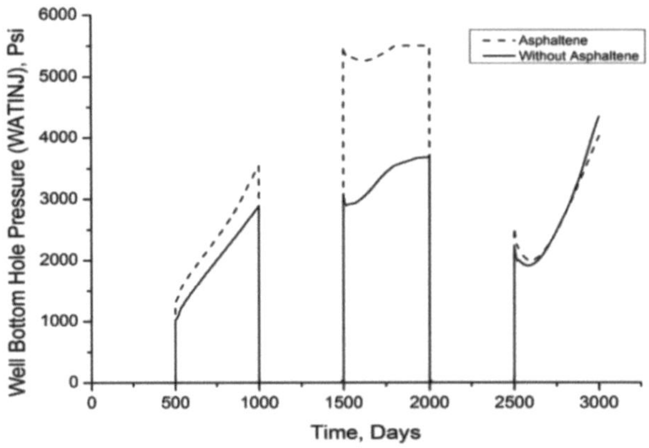

Figure 19.
Well bottomhole pressure history during water injection at the injector.

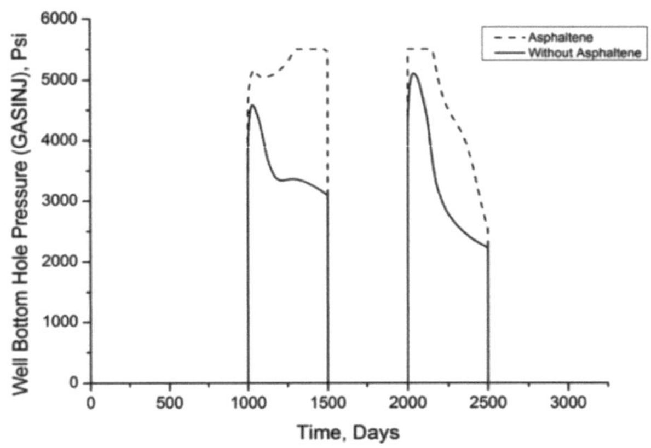

Figure 20.
Well bottomhole pressure history during CO_2 injection at the injection well.

oil production rate is obtained for the case without asphaltene. However, as can be seen, the asphaltene case can produce at lower rates compared to that case without asphaltene but in a longer production period, which has improved the ultimate oil recovery. Moreover, as can be seen in this figure, there are two pick points in field oil production curves for both cases. In each case, the field oil production increases during the gas injection but sharply decreases at the pick point. This can be explained in terms of early gas breakthrough time in the case without asphaltene and increasing gas-oil ratio in the production well, see **Figure 18**. The well bottomhole pressure values for water injection and gas injection wells are given in **Figures 19** and **20**, respectively. However, it should be mentioned that the production well in both cases is controlled with a bottomhole pressure mode of 500 psi. As shown in these figures, the asphaltene deposition has increased the well bottomhole pressure values. This can be caused by flow issues because of asphaltene deposition that caused some difficulty in terms of injectivity of these wells and in terms of porosity and absolute permeability reduction [34].

4. Conclusion and suggestions for future studies

From the outcomes of this investigation, we can infer the following conclusions:

- Asphaltene deposition alters rock wettability, increases water relative permeability at residual oil saturation, decreases oil relative permeability at irreducible water saturation, and changes the oil-water relative permeability curve cross-point. This suggests a shift from water-wet to more oil-wet or mixed-wet systems.

- In water-oil experiments, cumulative oil production decreases within two pore volumes due to the increased asphaltene deposition but ultimately increases over six pore volumes. Multiple mechanisms, including wettability alteration, surface film oil drainage, and interfacial tension changes, may contribute to improved oil recovery.

- Asphaltene deposition does not significantly affect gas-oil relative permeability and cumulative oil production in gas-oil systems.

- In three-phase systems, oil relative permeability values vary with asphaltene deposition until a certain gas saturation is reached, beyond which they converge, indicating minimal impact.

- Corey and LET correlations are the most suitable for history-matching relative permeability curves in water-oil and gas-oil systems.

- Nonlinear, multiregression analysis is used to establish correlations from the water and oil relative permeability experimental data based on asphaltene deposition. These correlations can be used in a numerical simulation study.

- The modeling and simulation of asphaltene processes during the WAG process in conventional simulators are explored. The field oil recovery factor with asphaltene deposition exceeds that without, though practicality remains uncertain.

- Asphaltene deposition increases well bottomhole pressure, potentially affecting flow assurance around wellbores.

- Experimental results suggest that varying weight factors for specific asphaltene deposition levels is necessary for accurate modeling, in contrast to the uniform weight factor currently used in conventional simulators.

Suggestions for future research:

- Subsequent research should focus on identifying the dominant mechanism, whether wettability alteration, surface film oil drainage, or changes in endpoints, contributing to enhanced oil recovery resulting from asphaltene deposition. Understanding these mechanisms in detail can provide valuable insights for optimizing EOR strategies.

- Evaluate the practicality and viability of injecting fluid volumes exceeding two pore volumes to achieve improved oil recovery in the presence of asphaltene precipitation and deposition. This exploration can guide decision-making in field operations and reservoir management.
- Enhance the accuracy of conventional compositional simulators by incorporating varying weight factors for specific levels of asphaltene deposition. This can lead to more realistic reservoir simulations and assist in predicting the impact of asphaltene on field production rates and reservoir performance.

Acknowledgements

The authors express their appreciation to the Carbon Capture & Utilization Department (CCUS) at PETRONAS Company, Malaysia, for generously allowing the publication of this work. Grateful acknowledgments are extended to Universiti Teknologi PETRONAS (UTP) for their indispensable support during this research. Sincere thanks go to Professor Mustafa Onur, McMan Endowed Chair Professor and Director of TU Petroleum Reservoir Exploitation Projects (TUPREP), for his invaluable assistance during my PhD as supervisor and thorough review of this book chapter. Special gratitude is also extended to Universiti Teknologi Malaysia (UTM) for providing the opportunity to collaborate on this publication. Parts of this chapter were previously published in the doctoral thesis by the same author [34].

Author details

Ahmad Khanifar[1] and Mansoor Zoveidavianpoor[2,3]*

1 PETRONAS Research Sdn. Bhd. (PRSB), Selangor, Malaysia

2 Faculty of Chemical and Energy Engineering, Department of Petroleum Engineering, Universiti Teknologi Malaysia, Malaysia

3 Malaysia Petroleum Resources Corporation (UTM-MPRC), Institute for Oil and Gas, Universiti Teknologi Malaysia, Johor, Malaysia

*Address all correspondence to: mansoor353@yahoo.com

References

[1] Kokal SL, Sayegh SG. Asphaltenes: The cholesterol of petroleum. In: Middle East Oil Show, Bahrain. 1995

[2] Nghiem LX, Coombe DA. Modelling asphaltene precipitation during primary depletion. SPE Journal. 1997

[3] Wang S, Civan F. Preventing asphaltene deposition in oil reservoirs by early water injection. In: SPE Production Operations Symposium, Oklahoma City, Oklahoma. 2005

[4] Mullins OC, Sheu EY, Hammami A, Marshall AG. Asphaltene, Heavy Oils, and Petroleomics. Springer; 2007

[5] Jamaluddin AKM, Creek J, Kabir CS, Mcfadden JD, D'Cruz D, Manakalathil J, et al. Laboratory techniques to measure thermodynamic asphaltene instability. Journal of Canadian Petroleum Technology. 2002

[6] Mullins OC. Review of the molecular structure and aggregation of asphaltenes and petroleomics. SPE Journal. 2008

[7] Khanifar A, Demiral B, Darman N. Study of asphaltene precipitation and deposition phenomenon during WAG application. In: Paper Presented at the 2011 SPE Enhanced Oil Recovery Conference, Kuala Lumpur, Malaysia. 2011. SPE-143488

[8] Sanchez NL. A general approach for asphaltene modeling. In: Latin American & Caribbean Petroleum Engineering Conference, Buenos Aires, Argentina. 2007

[9] Yi T, Fadili A, Ibrahim MN, Al-Matar BS. Modeling the effect of asphaltene on the development of the Marrat field. In: 8th European Formation Damage Conference, Scheveningen, the Netherlands. 2009

[10] Mansoori GA. Remediation of Asphaltene and Heavy Organic Deposits in Oil Wells and in Pipelines. USA: University of Illinois; 2010

[11] Negahban S, Joshi N, Jamaluddin AKM, Nighswander J. A systematic approach for experimental study of asphaltene deposition for an Abu Dhabi reservoir under WAG development plan. In: International Symposium on Oilfield Chemistry, Houston, Texas. 2003

[12] Rezaian A, Kordestany A, Jamialahmadi M, Moghadasi J, Khoshdaregi MY, Alipanah M, et al. Experimental and theoretical studies of flocculated asphaltene deposition from oil in porous media. In: Trinidad and Tobago Energy Resources Conference, Port of Spain, Trinidad. 2010

[13] Shedid SA. Influences of asphaltene deposition on rock/fluid properties of low permeability carbonate reservoirs. In: SPE Middle East Oil Show, Bahrain. 2001

[14] Wang S, Civan F. Productivity decline of vertical and horizontal Wells by asphaltene deposition in petroleum reservoirs. In: SPE International Symposium on Oilfield Chemistry, Houston, Texas. 2001

[15] Oskui G-RP, Jumaa MA, Abuhained W. Laboratory investigation of asphaltene precipitation problems during CO_2/hydrocarbon injection project for EOR application in Kuwaiti reservoirs. In: Kuwait International Petroleum Conference and Exhibition, Kuwait City, Kuwait. 2009

[16] Kalantari G, Moghadasi, Abdi R. Formation damage through asphaltene precipitation resulting from CO_2 gas

injection in Iranian carbonate reservoirs. SPE Production & Operations. 2008

[17] Minssieux L. Core damage from crude asphaltene deposition. In: International Symposium on Oilfield Chemistry, Houston, Texas. 1997

[18] Alizadeh N, Manshad AK, Fadili A, Leung E, Ashoori S. Simulating the permeability reduction due to asphaltene deposition in porous media. In: International Petroleum Technology Conference, Doha, Qatar. 2009

[19] Al-Maamari RSH, Buckley JS. Asphaltene precipitation and alteration of wetting: Can wettability change during oil production? In: SPE/DOE Improved Oil Recovery Symposium, Tulsa, Oklahoma. 2000

[20] Khanifar A, Abd Raub MR, Tewari RD, Zain ZM, Sedaralit MF. Experiment aided optimization of IWAG recovery process in a complex field with complex fluid: A case study. In: Paper Presented at the SPE Enhanced Oil Recovery Conference, Kuala Lumpur, Malaysia. 2015. SPE-154152-MS

[21] Khanifar A, Abd Raub MR, Tewari RD, Zain ZM, Sedaralit MF. Designing of successful immiscible water alternating gas (IWAG) coreflood experiment. In: Paper Presented at the International Petroleum Technology Conference, Doha, Qatar. 2015. IPTC-154152-MS

[22] Khanifar A, Abdul Manap AA. Maximizing oil recovery using a chemical application in offshore environment-case study. In: Paper Presented at the SPE Reservoir Characterization and Simulation Conference and Exhibition, Abu Dhabi, UAE. 2023. SPE-212618-MS

[23] Khanifar A, Subari IB, Raub MRA, Tewari RD, Sedaralit MF. The world's first offshore GASWAG EOR full field implementation. In: Paper Presented at the Abu Dhabi International Petroleum Exhibition & Conference, Abu Dhabi, UAE. 2021. SPE-208127-MS

[24] Majidaie S, Khanifar A, Onur M, Tan I. A simulation study of chemically enhanced water alternating gas (CWAG) injection. In: Paper Presented at the SPE EOR Conference at Oil and Gas West Asia, Muscat, Oman. 2012. SPE 154152

[25] Civan F. Formation damage mechanisms and their phenomenological modeling—an overview. In: European Formation Damage Conference, Scheveningen, the Netherlands. 2007

[26] Agbalaka CC, Dandekar AY, Patil SL, Khataniar S, Hemsath J. The effect of wettability on oil recovery: A review. In: SPE Asia Pacific Oil and Gas Conference and Exhibition, Perth, Australia. 2008

[27] Civan F. Reservoir Formation Damage. Houston, Texas: University of Oklahoma; 2000

[28] Kamath VA, Yang J, Sharma GD. Effect of asphaltene deposition on dynamic displacements of oil by water. In: SPE Western Regional Meeting, Anchorage, Alaska. 1993

[29] Honarpour M, Koederitz L, Harvey AH. Relative Permeability of Petroleum Reservoirs. Boca Raton, Florida: CRC Press; 1988

[30] Tarek A. Reservoir Engineering Handbook. 2nd ed. Gulf Professional Publishing; 2001

[31] Watt UH. Reservoir Simulation Handnotes. Scotland: Institute of Petroleum Engineering, Heriot-Watt University; 2008

[32] Johnson EF, Bossler DP, Naumann VO. Calculation of Relative Permeability from Displacement Experiments. Society of Petroleum Engineers of AIME; 1959

[33] Sendra. Sendra User Guide. Norway: Weatherford Petroleum Consultants; 2011

[34] Khanifar A. Investigation of the Effect of Asphaltene Presence on Relative Permeability Characteristics during Wag Process [Thesis]. Malaysia: University Technology Petronas; 2013

[35] Corey AT. The Interrelation between Gas and Oil Relative Permeabilities. Scinapse; 1954

[36] Lomeland F, Ebeltoft E, Thomas WH. A new versatile relative permeability correlation. In: International Symposium of the SCA, Abu Dhabi, UAE. 2008

[37] Burke NE, Hobbs RE, Kashou SF. Measurement and modelling of asphaltene precipitation (includes associated paper 23831). SPE Journal of Petroleum Technology. 1990

Chapter 3

Progress of CO_2 EOR and Storage Technology

Rui Wang, Qiaoyu Ge, Zengmin Lun and Maolei Cui

Abstract

CO_2 flooding is one of the most promising EOR technologies. The laboratory experiments have gradually evolved from early basic experiments to the revelation of physicochemical mechanisms and multi-scale physical simulation studies. The numerical simulation method moves toward the novel numerical simulation coupling compositional simulation with geochemical reaction and the stress field. Moreover, the optimization method starts to focus on the multi-objective optimization of CO_2 EOR and storage. Meanwhile, stratified gas injection processes and tools are crucial to implement balanced gas injection. The corrosion prevention technology is required to combine anti-corrosive materials and corrosion inhibitors. The injection-production adjustment is the priority to be considered in the early stage of gas injection. The chemical-assisted suppression methods, including foam agent, particles, and gel, are needed to implement in the later stage of gas injection. Gas channeling treatment with hierarchical management is crucial to tailor the different channeling channels. A full-chain optimization model based on net emission reduction and carbon footprint is suggested for the future system planning of CO_2 EOR and storage. In general, the next-generation CO_2 EOR technology not only aims to significantly increase the oil recovery but also achieves large-scale CO_2 storage, providing strong support for carbon neutrality goals.

Keywords: CO_2 EOR and storage, experiment, numerical simulation, multi-objective optimization, chemical-assisted method, full chain economic evaluation

1. Introduction

CO_2 is a type of gas with high solubility in both oil and water. As crude oil is dissolved with a large amount of CO_2, it will expand, with reduced viscosity and interfacial tension, forming a mixed-phase state with intermediate composition extracted. At the same time, the carbonic acid water formed by dissolving CO_2 in water can also improve rock wettability, acidify rocks, and improve the oil-water mobility ratio. Owing to the high performance of oil displacement, the technology of CO_2 flooding for enhanced oil recovery (CO_2 EOR) has been widely used. The CO_2 EOR process is typically operated in two ways: either water-alternating gas (WAG) or continuous CO_2 injection. WAG flooding typically employs the use of a relatively large, initial slug of CO_2 to mobilize residual oil. **Figure 1** depicts the cross section of a typical

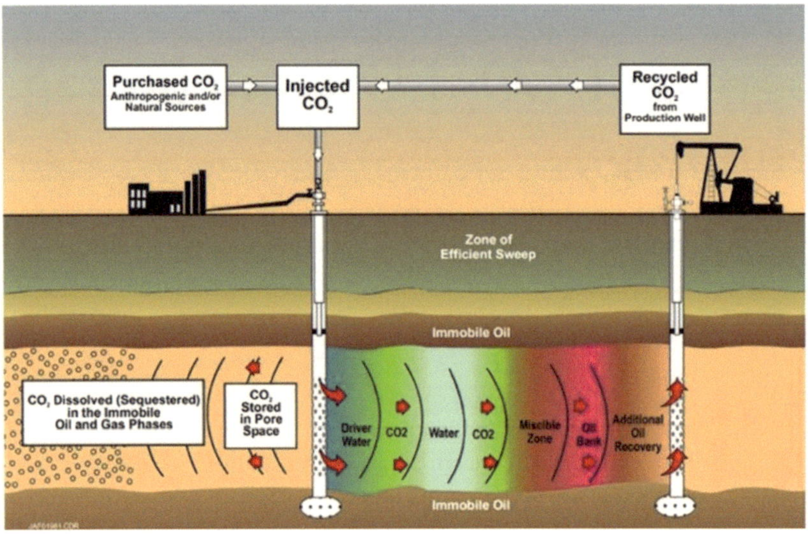

Figure 1.
Cross section of a typical WAG CO_2-EOR operation. Source: Advanced Resources International.

WAG CO_2 EOR operation [1]. Nevertheless, in continuous CO_2 injection, the gas is injected continuously without water slugs for the distribution.

As a predominant candidate for EOR application, CO_2 has various advantages owing to its super-critical status at reservoir conditions. Significant viscosity reduction can be achieved as CO_2 dissolves into crude oil. The higher the viscosity of crude oil, the greater the degree of viscosity reduction, and the higher the improvement in oil mobility. Meanwhile, the dissolution of CO_2 induces volumetric expansion of crude oil and increases the formation energy. As CO_2 gets in contact with fresh oil, mass transfer happens along with the dissolution, extracting a certain amount of light hydrocarbon components from the crude oil. Experimental results show that the components between C_2 and C_{20} could be extracted completely as long as the pressure is sufficiently high, which reduces the interfacial tension significantly. Furthermore, the miscibility state could be achieved as the miscible zone is being formed. The miscible zone is closely related to the extraction and the condensation between CO_2 and crude oil. Besides, the diffusion effect of CO_2 in crude oil and brine is also one of the most important EOR mechanisms. The mineral dissolution is another impact as CO_2-water reacts with carbonate minerals. In a word, the EOR mechanisms of CO_2 flooding are remarkable [2–4].

CO_2 EOR technology has been developed since the 1950s. The first patent for applying CO_2 flooding in EOR was proposed in 1952. A lot of research on the CO_2 EOR mechanism was performed later. In the 1970s, the first commercial CO_2 EOR field application was carried out in the SACROC oilfield in the United States. In the 1980s, the industrialized application of CO_2 EOR was promoted in the Permian Basin, forming a CO_2 EOR-supporting technology system. Since the 1990s, attentions have been paid to the improvement on sweep efficiency of CO_2 EOR. Research was conducted on well pattern matching and chemical agent for channeling suppression, resulting in the development of various technologies, such as CO_2 vertical well injection with horizontal well extraction, and CO_2 injection-assisted gravity drainage.

In 2010, the U.S. Department of Energy released a report outlooking the research directions of next-generation CO_2 EOR technology. The two main research focuses are CO_2 EOR and storage, and tight unconventional oil and gas development. The CO_2 injection slug would be greatly increased to more than 0.8 ~ 1.0 HCPV. More attention will be paid to the matching of well pattern and well type, and chemical-assisted efficiency enhancement. The integrated optimization technology of CO_2 EOR and storage is becoming increasingly important. In addition, the application of CO_2 EOR technology has gradually shifted from conventional reservoirs to new fields such as tight unconventional reservoirs and residual oil resources. Meanwhile, the integration of CO_2 EOR with hydraulic fracturing techniques is also taking place, which is becoming a new focus. By the end of 2020, the number of CO_2 EOR projects has been increased to 142 in the U.S. The updated survey shows that incremental oil recovery from CO_2 EOR in the U.S. was approximately 273,000 barrels per day (bpd) in 2020. Although it indicates a decline of about 9% from the 2019 survey (total of 299,000 bpd), CO_2 EOR remains an attractive option for generating revenue as the 45Q tax incentive for anthropogenic CO_2 is finalized [5–9].

As a technology chain, CO_2 EOR is a systematic project, mainly including experiments, numerical simulation, reservoir engineering, injection and production technology, anti-channeling and sealing, economic evaluation, etc. These technologies will be described in detail below.

2. Experimental evaluation technology of CO_2 EOR and storage

Experiments are the most direct way to study the mechanism of CO_2 EOR. By introducing high-temperature and high-pressure conditions, we can realistically simulate the interaction process of CO_2 with oil, water, and rock under reservoir conditions, therefore revealing the physicochemical nature of CO_2 EOR.

The development of CO_2 EOR experiments has gone through three main stages (**Figure 2**). The first stage is the basic evaluation experiment, including the PVT phase experiment, the long slim tube experiment, and the long core displacement experiment. The PVT phase experiment is to study the changes in reservoir fluid parameters before and after gas injection, which then provides basic parameters for reservoir numerical simulation study and reservoir engineering scheme. The long slim tube experiment is designed to determine the minimum miscible pressure between injected gas and crude oil, and thus to quickly evaluate the feasibility of gas flooding. The long core displacement experiment is a necessary means to simulate the effect of gas flooding under reservoir conditions, which helps to optimize the injection parameters and injection methods in the gas flooding process.

The second is the mechanistic study stage, which reveals the physicochemical nature of the CO_2 EOR process, such as the multiphase oil-gas-water interfacial characteristics, the diffusion and the extraction mechanisms, the solid phase deposition, and the multiphase seepage mechanism. The in-depth theoretical study of the mechanism provides a fundamental understanding of the technical process of CO_2 EOR for researchers. Moreover, the CO_2 storage mechanisms are new focuses. The dissolution of CO_2 into brine induces the loss of the injected gas volume. The formation of carbonate water causes the mineral dissolution and sequestrates some parts of CO_2 underground.

The third stage is the physical simulation stage. Multi-scale physical simulation with physical similarity is carried out to restore the conditions of CO_2 flooding in the reservoir. This technology mainly presents the following characteristics: “small, precise, and large.” “Small” means the small pore-scale in the physical models. The physicochemical nature of nano-pore-scale conditions is gradually revealed, which provides theoretical

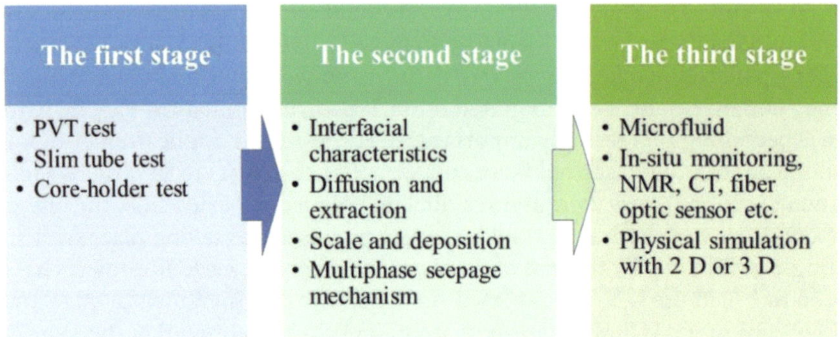

Figure 2.
The development history of experiments on CO_2 EOR and storage.

support for the knowledge and understanding of the oil displacement mechanisms. "Precise" is achieved through the characteristics of automation, high resolution, in situ observation, etc. Accurate data can be obtained by characterizing the experimental process through in situ NMR or CT, new fiber optic sensors, automated robots, and other advanced means. "Large" means the physical similarity of the physical models by scaling up to reservoir conditions. At the same time, various new sensors have been integrated to enable real-time measurement of different field parameters during the displacement process and monitoring of gas flooding leading edge and saturation changes [10–13].

To sum up, the experimental technologies of CO_2 EOR have evolved from the initial determination of field-based parameters, through the interpretation of field problems, and restoration of production processes, to new digital experimental methods based on digital twins. In the future, CO_2 EOR indoor experimental technology will develop toward more realistic and credible, more instructive, and more flexible and diverse. In addition, it gradually breaks through the original technical limitations, and changes from focusing on oil displacement to focusing on the sequestration effect, that is, focusing on the retention mechanism of CO_2 itself in the reservoir. Experimental technology has also developed from formation physics to get integrated with geochemistry, petrophysics, mechanical engineering, information science, and other disciplines, forming a new multidisciplinary field.

3. Numerical simulation technology of CO_2 EOR and storage

Numerical simulation is the basic means to study the engineering application of CO_2 EOR. In 1953, Bruce G.H and Peaceman D.W simulated one-dimensional gas-phase unsteady radial and linear flows [14]. Later, it was extended to reservoir numerical simulation, that is, using a computer to solve the mathematical model of the reservoir, simulate the subsurface oil and water flow, provide the oil and water distribution at a certain time, and thus predict the reservoir dynamics.

3.1 Numerical simulation models

The current numerical models of CO_2 EOR include five main types: CO_2 EOR based on the black oil model, the transport-diffusion model, the pseudo-component model, the full-component model, and the novel component model.

The first model adopts the control equations of the black oil model and approximates the viscosity relationship with the mixing parameters, which can simulate various fingering phenomena and the effect of fingering on the areal sweep efficiency. The advantages of the model are simpler, less computational effort, higher stability, and better matching the actual miscible flooding process. The disadvantage is that it can only approximate the CO_2 EOR process with poor accuracy. That is, it cannot fully reflect the transport-diffusion process and mechanisms.

In the transport-diffusion model, the fluid is divided into two components: oil and solvent. This model considers the strong diffusion effect, which makes it adaptable to the process of miscible flooding. However, the model does not adequately describe and modify the fluid properties, and cannot truly reflect the inter-phase and inter-component changes of the fluid in the CO_2 flooding process. And, there is severe numerical dispersion when modeling the displacement leading edge.

The pseudo-component model, developed after the transport-diffusion model for component calculations, is similar to the pure-component model. The only difference is that the iteration of the equilibrium constant K is simplified. That is, the value of K is given a pressure-, temperature-, and component-dependent equation without participating in the iterative correction. The K value varies with the model parameters and requires a small computational effort, making the K value equation suitable for fast computation through experimental regression. The disadvantage is that the description of the components is not very accurate because the iterative solution of the fugacity equation and the gas-liquid equilibrium constants are not performed.

The full-component model can better simulate the mass transfer and component changes, including the simulation of gasification, condensation, expansion, and other processes. The disadvantage is that the processes of CO_2 EOR and the multi-staged CO_2 extraction of crude oil are not well considered. At the same time, correction of parameters such as fluid density and viscosity is not incorporated.

The novel component model of CO_2 EOR has the advantages of both the component model and the black oil model, which can improve the calculation speed based on the simulation of mass transfer and component changes. By involving various practical conditions, it can simulate the solubility of CO_2, the compressibility of rocks and fluids, the heterogeneity and anisotropy of the formation, the gravity effect of fluids, and the correction of fluid density, viscosity, solubility, and relative permeability, etc. [15–20].

3.2 Synchronous simulation of CO_2 EOR and storage

With the increasing importance of CO_2 emission reduction, the effectiveness of CO_2 sequestration has gradually attracted attention. The consideration of storage mechanisms in the CO_2 EOR process is the latest research direction in the development of component numerical simulation technology. At present, the Eclipse and CMG numerical simulation software is commonly used to perform the phase calculation of CO_2 in both oil and gas phases by flash calculation. However, for the conditions where CO_2, oil, water, and gas co-exist, the dissolution of CO_2 in formation water cannot be considered in the models. In addition, for the latest sequestration models or simulators, the geochemical reaction process is considered, while the effects of porosity and permeability parameters caused by geochemical reactions are not readily available. Due to the computationally intensive nature of the module considering the geochemical reaction, it is not possible to synchronize its operation with the oil displacement module, and it can hardly achieve the integration of CO_2 EOR and storage. Synchronous simulation of CO_2 EOR and storage requires simultaneous consideration of CO_2 dissolution in both oil and water phases, chemical reactions between carbonate, water, and reservoir minerals, and other storage mechanisms. The commonly adopted approach is to address the effect of CO_2-water reaction with rock minerals on reservoir porosity/permeability parameters through an equivalence treatment. As shown in **Figure 3**, by invoking the equations of porosity/permeability changing with carbonated water injection acquired in the laboratory, the program is figured out and used to automatically analyze the calculation results. The results are further used to calculate the effect of the reservoir mineralization reaction on the porosity and permeability of each grid during that simulation time period. Then, Eclipse is restarted to calculate the next time period. The cycle is repeated until the whole calculation is completed. In addition, the partitioning effect of CO_2 in oil and water is mainly considered by the interpolation of CO_2 solubility tables. That is, the equilibrium process of CO_2 in oil and gas phases is still calculated through the flash equation, while the equilibrium process of CO_2 in water is calculated using the gas-water solubility table interpolation. Besides, the capillary hysteresis is considered by introducing the parameter of relative permeability hysteresis effect on the basis of fitting experimental data. The above method enables the equivalent consideration of the sequestration mechanism in the CO_2 EOR process, thus achieving simultaneous numerical simulation of CO_2 EOR and storage processes [21].

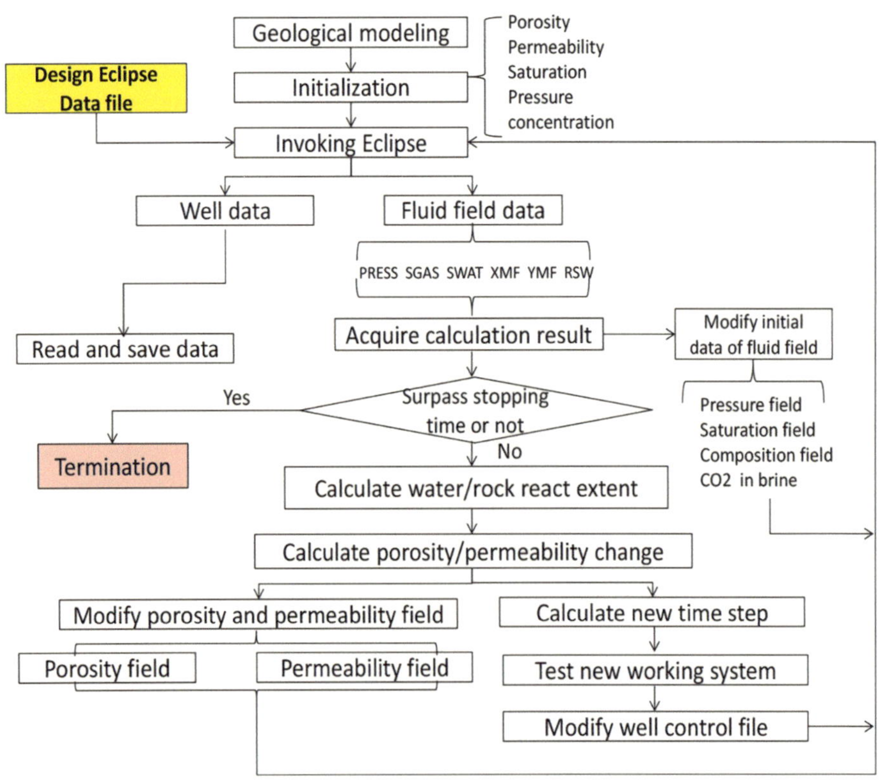

Figure 3.
Numerical simulation flow diagram for CO_2 EOR and storage.

4. Optimization design method for CO_2 EOR and storage

Optimization design for CO_2 EOR and storage is a key step in engineering practice, which has evolved from the initial dimensionless curve method, to the streamlined simulation evaluation, the single objective optimization, and the current multi-objective optimization and system decision-making. Originally, only the oil displacement process was optimized. Later, the optimization method developed into the integrated CO_2 EOR and storage optimization, which provides technical means for large-scale CO_2 storage during the process of oil displacement.

4.1 Water/gas flooding characteristic curves

The early dimensionless curve method was used to predict engineering implementation according to the empirical method before the advent of computers. The production parameter characterization was carried out according to the water flooding characteristic curve, which provides a basis for decision-making in the prediction of production parameters. The so-called water flooding characteristic curve is a relationship between cumulative oil production, cumulative water production, and accumulation of fluid production during water injection. These curves have been widely used for water injection development of dynamic and recoverable reserves forecasts. Based on actual oilfield production data, water flooding characteristic curves are used to predict field production data and recovered reserves in a statistical way. Owing to the intuitive data and easy calculations, this method is commonly used by field reservoir engineers. Currently, there are 10 types of water flooding characteristic curves. The most commonly used ones are type A, type B, type C, and type D. These characteristic curves are the relationship initially proposed by former Soviet Union scientists through statistics on actual production data and laboratory data aiming to characterize water flooding performance. Later on, the curves were extended by researchers and derived into new expressions, which were named after the researchers. In 1978, Tong Xianzhang named the Maksimov water flooding characteristic curve as type A curve. The three other water flooding characteristic curves were named as type B, type C, and type D curves sequentially by Chinese researchers [22–25].

However, no standard gas flooding characteristic curve has been developed for CO_2 EOR. The main challenge is the complicated relative permeability characteristics of the three phases: oil, gas, and water. Due to the intense physical and chemical interaction process among CO_2, oil, and brine, the physical parameters of the fluids take on the dynamic change, especially for crude oil and CO_2. Moreover, the flow mechanism becomes more complex, which makes it difficult to describe the process in the theory of water flooding characteristic curve. However, the segmented log-log correlations between the cumulative gas production and the cumulative oil production have been established through theoretical derivation, forming a method to describe the characteristic curve of gas flooding in the process of CO_2 immiscible displacement in low-permeability reservoirs (**Figure 4**). Combining with indoor experiments and field tests, the empirical formula has been verified. It shows that the cumulative gas production and cumulative oil production present a segmented double-logarithmic linear variation relationship, which clarifies the rationality of the characteristic curve of gas flooding [26]. The responding performance of gas flooding could be predicted based on the gas flooding characteristic curve described above. The gas breakthrough point could be determined exactly. The incremental oil recovery could be also calculated theoretically.

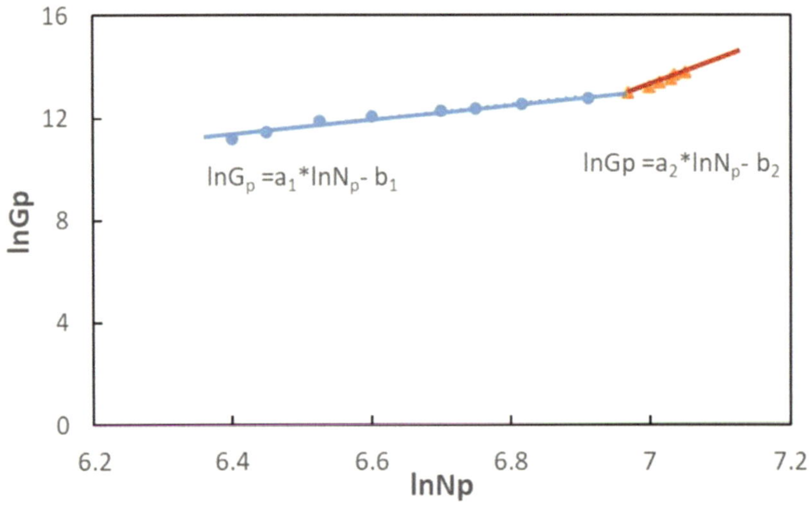

Figure 4.
Gas flooding characteristic curve for CO_2 immiscible flooding.

4.2 Synchronous optimization of CO_2 EOR and storage

With the development of numerical simulation, the optimization design of reservoir engineering has gradually shifted to multi-parameter from single-objective optimization. For CO_2 EOR, the optimization design of various injection and production parameters by the incremental recovery factor or oil exchange ratio provides the optimal objective for reservoir engineering plan development. However, for CO_2 EOR and storage, there are two objects to be optimized. Since there are differences or even contradictions in the design concepts between conventional CO_2 EOR and CO_2 storage technologies, it is crucial to find commonalities between the two and form an integrated optimization technique for both CO_2 EOR and CO_2 storage. For CO_2 EOR, the gas source comes from natural gas or industrial gas. The injection method is continuous injection or WAG injection during secondary or tertiary oil recovery. The common injection volume is about 0.25 ~ 0.30 HCPV. The optimization goal is the optimal recovery factor, which requires injecting as little CO_2 as possible to produce more oil. For CO_2 storage, however, the gas source is limited to industrial gas. The timing of injection takes place early or after the reservoir is abandoned. The injection volume is often very large, typically in the range of 1.0 to 1.5 HCPV, which requires the injection of as much CO_2 as possible to produce the most oil and water. The optimization target is a high storage ratio. In the project of integrated CO_2 EOR and storage, it is necessary to use parameters that can describe both the oil recovery factor and the CO_2 storage ratio and evaluate the project comprehensively. Therefore, in order to adapt to different decision scenarios and engineering objectives, a comprehensive evaluation function of the integrated oil displacement-embedded effect is introduced and used as the evaluation criterion to study the optimal injection and recovery plan under different scenarios [15, 27, 28].

$$f = \omega_1 \frac{M_o}{M_{OIP}} + \omega_2 \frac{M_{CO_2} / \rho_{CO_2}^{R,0}}{PV} \tag{1}$$

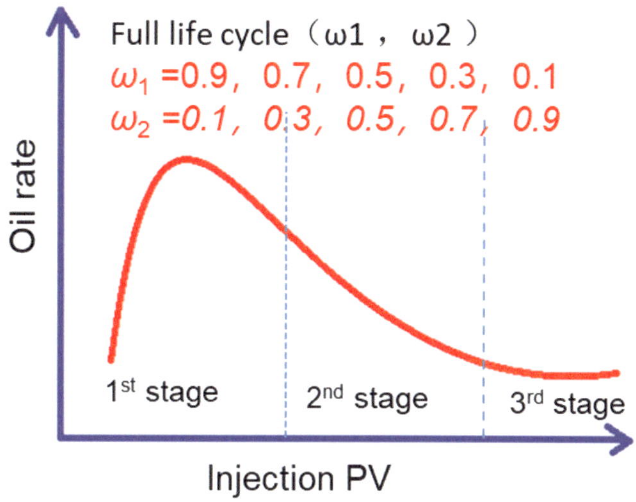

Figure 5.
Global optimization method for CO_2 EOR and storage.

where ω_1 and ω_2 are the weights of the CO_2 EOR term and the CO_2 storage term, respectively, and M_{OIP} is the volume of oil reserve at the beginning of optimization. $\rho_{CO_2}^{R,o}$ is the average density of all gas-phase CO_2 in the reservoir. PV is the pore volume of the reservoir. In short, f is a harmonic value that measures the sum of the proportion of crude oil recovered by CO_2 EOR and the proportion of the original pore space of the reservoir occupied by the injected CO_2.

In the process of synchronous optimization of CO_2 EOR and storage, the importance of CO_2 EOR and CO_2 storage under different scenarios is characterized by adjusting the ratio of ω_1, ω_2. A reasonable selection of weights is very important for the optimization process. If the primary objective is to improve the recovery of crude oil, ω_1 can be taken as 1. If the primary objective is to maximize the storage of CO_2, ω_2 can be set as 1. There is no unified evaluation criterion for CO_2 storage in China as the legal and taxation policies for CO_2 storage are unknown. As shown in **Figure 5**, during the full life cycle of CO_2 EOR and storage projects, oil recovery is often the main focus at the beginning. Once the displacement effect reaches its limit and the gas-to-oil ratio (GOR) increases significantly, the CO_2 EOR and the CO_2 storage are both important. At the later production stage when there is almost no continuous crude oil production, more attention is paid to the storage amount. In the optimization study of CO_2 EOR and storage, there is only one optimal objective and a set of parameters to be optimized at each stage, so the best-optimized parameter values can be determined by the extreme or inflection points of the simulation results [21].

5. Supporting engineering technologies of CO_2 EOR and storage

The injection and production engineering is the key component of CO_2 EOR technology, which plays an important role in carrying forward and downward. It is the guarantee for completing the development targets proposed in the reservoir development plan, as well as the basis and starting point for the construction of surface engineering. Researchers have done a lot of work on CO_2 injection, oil production, and anti-corrosion technologies, with impressive achievements, which greatly promote the development of this technology [29–34].

5.1 CO_2 injection technology

5.1.1 Blanket gas injection technology

At present, the blanket gas injection technology is more commonly used in CO_2 EOR field tests. For reservoirs with a relatively uniform suction profile, the blanket gas injection works well for an effective displacement of multiple oil reservoirs. The traditional injection string adopts an integral string design, which exposes various problems: poor sealing effect of the packer, fast rise of casing pressure, short oil-casing pressure balance time, etc. In addition, the formation may be damaged by the necessary well operation, such as bleeding, and well-killing. The operation under pressure is expensive as well. To avoid these problems, the no-kill gas injection pipe column with a split-body releasing tool is proposed, which works with the functions of anchoring, backwashing, split-body releasing tool, and operation without killing the well. The corresponding string-supporting tools are also incorporated.

5.1.2 Stratified gas injection technology

For multi-layered reservoirs with large differences between layers, gas channeling and flow control problems caused by the viscosity difference between CO_2 and crude oil and the strong heterogeneity are key issues in the development process of CO_2 injection. Gas channeling will cause the injected CO_2 to form an ineffective cycle, which greatly reduces the CO_2 sweep volume and the oil recovery factor. Meanwhile, the CO_2 will cause serious corrosion problems after gas channeling. Compared to blanket gas injection, stratified gas injection is more helpful in achieving uniform reservoir utilization and avoiding gas channeling caused by reservoir heterogeneity.

Commonly used stratified gas injection technologies include single-tube and concentric double-tube stratified gas injection. For the single-tube stratified injection technology, a packer is used to separate the injection layers. Each injection layer corresponds to a distributor, which can be opened and closed to achieve the alternating injection between different layers. The downhole stratified injection volume is controlled by adjusting the dispenser nozzle. In fact, the CO_2 EOR single-tube stratified gas injection technology involves several problems. First, the downhole stratified gas injection tools and process design require further optimization. The size of the injection nozzle is small, which can be easily blocked due to asphaltene precipitation in the injection well. In addition, measurement of CO_2 at a supercritical state is difficult, and the stratified gas injection measurement and testing are challenging due to complicated nozzle flow characteristics. Moreover, the CO_2 injection pressure is high, leading to high safety risks during the test.

For the concentric dual-tube stratified gas injection technology, the upper oil layer is injected through the outer and central tubing annulus, and the lower oil layer is injected through the central tubing. Limited by the size of the casing in the old wells, the tubing may be blocked during gas injection in concentric dual-tube. Besides, well operation in dual-tube is difficult and expensive. Currently, the stratified gas injection technology is still in the R&D and testing stage, which requires further research and improvement.

5.1.3 Gas injection equipment

Considering gas injection sealing, corrosion prevention, and operational requirements, stainless steel injection wellheads and airtight tubing strings for gas injection are utilized. Different types of injection tubing strings are developed, including no-kill well injection tubing strings, mechanically anchored injection tubing strings, and self-balancing injection tubing strings. These tubing strings have a maintenance-free period of more than 27 months. For the purpose of balanced gas injection, the CO_2 eccentric injection tubing string and the support compensation self-balancing injection tubing string are designed and developed to achieve CO_2 stratified gas injection.

For the problems of serious corrosion and increasing GOR in the production wells with ordinary carbon steel wellheads and 3Cr or 13Cr stainless steel downhole tools, inserted oil recovery tubing strings and multi-functional oil recovery tubing strings are developed. The pumping efficiency of oil recovery pumps is improved by employing measures such as an anti-corrosion pump + high-efficiency gas anchor, spiral flow sieve tube + anti-gas pump, over-bridge pump + long tail pipe, which helps to extend the maintenance-free period of oil wells as well. In addition, the production efficiency is also improved by the combination of various other measures. A combined capillary tube pressure measuring device and corrosion tube piece are used to monitor the real-time gas flooding efficiency and corrosion state. A corrosion inhibitor is added to the annulus to protect the casing. Production control valves are added to allow operation without well-killing.

5.1.4 Gas injection station for different gas supply

In general, different injection technologies have been developed according to the characteristics of the CO_2 feed source. A pressurized injection station is employed for gas injection in the case of large-scale and continuous gas supply. This injection station incorporates the booster unit that boosts, heats, and distributes to the dispenser room, and the injection unit from the dispenser room to the single well. An integrated dual-media distribution process with alternative gas and water injection is established. The liquid CO_2 pump injection technology is developed with self-pressurized CO_2 storage tanks. For discontinuous gas supply, a convenient and flexible skid-mounted injection method is adopted, which integrates an injection system, automatic control system, and heating system, to meet the injection needs of different geological conditions, scales, and pressures.

5.2 Oil production technology

Oil production technology in China mainly relies on conventional oil pumps. As the field test progresses, the oil production wells face the problems of elevated casing pressure, elevated GOR, and intermittent gas production, which seriously affect the pumping efficiency and production of the wells. As the CO_2 EOR test is performed in fields with low permeability and low production, gas-liquid separation is difficult. Conventional lifting methods used in wells with large GORs have poor pumping liquid filling, low pumping efficiency, and an "air lock" problem. This makes the pump unable to work normally and unable to maintain the CO_2 EOR effectiveness. At the same time, "liquid surface shock" may occur, which accelerates the damage to the downhole equipment, such as the pumping rod, valve stem, valve cover, pump valve, oil pipe.

5.2.1 Elevated casing pressure

For the issue of rising casing pressure, gas energy in the casing can be utilized to help the rod pump lift the fluid. Combining the rod pump and gas lift oil production techniques, a gas lift valve is installed at a certain depth of the well to help the rod pump lift the fluid. During the pumping process, when the annular casing pressure rises above the opening pressure of the gas lift valve, the gas enters the tubing through the gas lift valve to reduce the density of the fluid in the tubing and hence lift the fluid. When the annular casing pressure is lower than the opening pressure of the gas lift valve, the gas lift valve closes. By installing the gas lift valve, the CO_2 gas energy of the breakthrough can be effectively used, and the casing pressure can be automatically controlled within a lower range so that the system is maintained at a dynamic equilibrium state.

5.2.2 Elevated GOR

For the issue of high GOR, we can increase the inlet pump pressure by controlling the bottom hole pressure, meanwhile installing a gas-liquid separator under the pump. When the produced fluid enters the wellbore, the gas flows upward due to gravity, and the liquid flows downward into the tubing string to achieve preliminary separation. When the liquid enters the gas-liquid separator after gravity separation, it is in fact a mixture of oil-liquid-gas-sand. The mixture enters from the lower inlet and travels down along the spiral surface. The denser sand particles travel along the outside of the spiral under the centrifugal force and enter the tail pipe of sedimentation. The gas and liquid travel down along the inside and enter the inner pipe from the bottom, and then enter the spiral gas-liquid separator, where the liquid travels upward along the outer spiral and gets discharged through the drainage sieve tube, entering the oil pump sieve tube. The gas goes up along the inner surface into the gas collection hood and gets discharged by the exhaust valve, entering the casing-tubing annulus. When the gas affects the lift due to increasing gas-liquid ratio, the use of gas anchors and anti-gas pumps can be considered to achieve a better lifting effect. The anti-gas oil pump provides a channel for the gas in the pump by setting a hollow pipe. This increases the liquid filling factor in the working cylinder, reduces the GOR in the pump, eliminates the interference of gas, prevents gas lock, and effectively improves the pumping efficiency.

5.2.3 Produced gas recovery

In addition, for the problem of gas reinjection during the CO_2 EOR process, there are mainly three different produced gas recovery technologies. The first is the CO_2 flooding gas recovery and separation technology applicable to large-scale, low-to-medium CO_2 content distillation coupled with low-temperature fractionation. This technique generally adopts chemical absorption decarbonization technology. The second is the low-temperature fractionation decarbonization technology applicable to large-scale, high CO_2 content cases. The third one is suitable for small-scale skid-mounted produced gas reinjection systems with a CO_2 capture rate greater than 80% and purity greater than 95%. In addition, the following three methods can be applied for the treatment of produced gas: direct reinjection, CO_2 dilution reinjection, and rectification-distillation reinjection. If the light hydrocarbon content in the produced gas is high, the rectification-distillation recovery technique can be used.

After the light hydrocarbon recovery, the remaining gas can be directly reinjected. If the CO_2 content in the produced gas is low, it can be reinjected by mixing with pure CO_2 to increase the CO_2 concentration. If the CO_2 content is high, it can be reinjected directly [35–37].

5.3 Anti-corrosion technology

Although dry CO_2 gas is not corrosive, serious corrosion of carbon steel can be caused when CO_2 is dissolved in water. Typical features of CO_2 corrosion include local pitting, moss-like, and table-top corrosion, which often cause accidents such as tubular perforation, fracture, gas leakage, and wellhead damage, thus affecting normal production. Since most of the CO_2-flooded gas injection wells and oil production wells are old wells, the wellbore was designed and completed as a conventional oil well, where carbon steel combined with J55, N80, and P110, was used for the casing, which is seriously corroded. According to the site anti-corrosion requirements, an indoor high-temperature and high-pressure reactor was used to simulate the site CO_2-containing environment, and the effects of the tubing, CO_2 partial pressure, temperature, flow rate, solution mineralization, pH value, and other factors on corrosion were analyzed and evaluated. The results show that the corrosion rate of 13Cr material is low, basically about 0.01 mm/a, while the corrosion rate of common carbon steel material is as high as 5–7 mm/a. Therefore, anti-corrosion measures must be taken. Considering the characteristics of injection and production processes in the CO_2 test area, the following measures can be taken to prevent or delay the corrosion of CO_2 on the tubular: optimizing the tubing string structure, utilizing corrosion-resistant materials, and injecting corrosion inhibitors.

5.3.1 Tubing string structure optimization

For injection wells, the annulus above the packer can be filled with oil-based annulus protection fluid to eliminate stress corrosion environment and avoid corrosion of carbon steel casing and tubing. When implementing water-gas alternation in injection wells, to prevent contact between CO_2 and water, gas injection wells must be filled with a corrosion inhibitor plug before water-gas alternation. For production wells, based on the characteristics of the oil production process, combined with research results of corrosion and protection technology, corrosion-resistant materials and corrosion inhibitors can be employed locally to prevent or delay the tubular corrosion. CC-grade anti-corrosion wellheads and coated or liner tubing are used for the production. Pumping rods, pumps, plungers, vanes, ball seats, and other components need to resist CO_2 corrosion as well. For wells without packers, a corrosion inhibitor is added from the casing-tubing annulus. For wells with packers, a corrosion inhibitor is added in the annulus above the packer. If the downhole tools are made of corrosion-resistant materials and the tubing is coated or lined, protection of the annular casing also requires the addition of corrosion inhibitors. Corrosion test pegs or rings are installed at representative locations to regularly monitor the corrosion of the tubing in the well and to optimize the corrosion protection program.

5.3.2 Corrosion-resistant alloy

According to the current development of related technology in China, the anti-corrosion technologies can be divided into two aspects:

anti-corrosion-technology-controlled and corrosion-resistant-alloy-controlled. The anti-corrosion technology includes corrosion inhibitor anti-corrosion, and coating anti-corrosion, which can be selected according to the actual situation in the region. Among all the common anti-corrosion measures, internal coating corrosion and cathodic protection corrosion are limited in use as the application conditions are relatively harsh. The measures of ordinary carbon steel with corrosion inhibitors and corrosion-resistant alloys are used more frequently. Although the initial investment for using corrosion-resistant alloy is higher than other measures, the corrosion protection performance is better. The ordinary carbon steel assisted with corrosion inhibitor costs less initially, while the later operation process is more complex, which leads to higher capital investment. Particularly for the wells with high temperature and high pressure, corrosion-resistant alloy works much better than the normal carbon steel.

5.3.3 Corrosion inhibitor

In projects where CO_2 injection time is relatively short, ordinary carbon steel with corrosion inhibitor is more practical for corrosion protection. During the corrosion inhibitor injection, the formulations of the corrosion inhibitor are critical. It developed from the initial imidazoline-type CO_2 corrosion inhibitor into an agent system that integrated scale inhibition, sterilization, and corrosion prevention, which can effectively inhibit SRB bacterial corrosion and scaling. When applied in the field, it is also necessary to carry out corrosion inhibitor formulation evaluation in combination with the reservoir conditions in the test area to determine the optimal corrosion inhibitor formulation. At the same time, corrosion inhibitor dosing concentration, dosing method, and dosing cycle are the main factors affecting the corrosion protection effect. At present, the research on the mechanism of CO_2 corrosion is getting in-depth, and the corrosion protection measures have achieved certain progress. The key problem is how to take the simplest, most effective, and most economical anti-corrosion measures considering the complex factors affecting the site.

5.3.4 Cathodic protection

As the partial pressure of CO_2 rises, the amount of CO_2 dissolved in water rises and the pH of the aqueous solution decreases, which greatly increases the possibility of the cathodic depolarization process of hydrogen, thus causing the corrosion to intensify. Therefore, cathodic protection can also help in corrosion protection. However, the operation process of cathodic protection is complex, the construction program is susceptible to the site environment. This makes the implementation of cathodic protection expensive and the application is thus limited. Coating anti-corrosion treatment is another common anti-corrosion measure. As the coating is performed at ultra-high temperature, the quality of the equipment can be improved, so that it is not affected by CO_2 corrosion. However, the application of coating at the wire buckle is challenging, which makes it not applicable to the cases under loading.

The integrity of the wellbore plays an important role in the implementation of CO_2 EOR technology, for both oil displacement and CO_2 storage processes. Anticorrosion technology is critical to wellbore integrity to ensure proper implementation of the CO_2 EOR process and to establish a solid foundation for long-term safe storage of CO_2.

6. Research on gas channeling suppression system and technology of CO_2 EOR

So far, various methods have been applied to prevent CO_2 gas channeling, including mechanical plugging, injection and production parameter adjustment, WAG injection, and chemical-assisted methods. By using mechanical plugging, injection rate control, and WAG, we can improve the CO_2 distribution in the injected well section and increase the vertical sweeping of injected CO_2. However, it is only effective to adjust conformance in the vicinity of the wellbore but not for formations far away from the injection well. When CO_2 passes through the perforation holes, the flow is controlled only by the reservoir heterogeneity and the flow mechanisms such as channel and gravity overburden. By injecting water and CO_2 alternatively, we can enhance both microscopic oil displacement efficiency and macroscopic sweep efficiency of CO_2 injection. This can prevent viscous fingering and delay the early gas breakthrough.

Chemical-assisted CO_2 EOR mobility control or gas channeling suppression technology is one of the main development directions of the next-generation CO_2 EOR technology identified by the U.S. Department of Energy. CO_2 EOR mobility control and sealing methods mainly involve the injection of organic or inorganic chemical agents to increase CO_2 viscosity or reduce effective permeability, thus reducing the mobility ratio and improving the oil displacement effect. First, CO_2 thickening agents are used to thicken CO_2 by adding polymer compounds such as siloxane polymers, fluoropolymers, and polymethacrylates to the injected gas. It can generally increase CO_2 viscosity by 20–30 times. The disadvantages are the low solubility of polymer compounds in CO_2 and the high cost. In addition, an inorganic gel system is another way to control CO_2 mobility. The reaction between CO_2 and silicate is used to form inorganic silicate gels. The advantage of this method is its lower cost, while the disadvantage is that the time of gel formation is difficult to control.

In the technique of organic ammonia salt profile control, CO_2 is used to react with organic amines to form a salt containing crystalline water to plug the formation. Commonly used amines include monoethanolamine (MEA), diethanolamine (DEA), triethanolamine (TEA), N-methyldiethanolamine (MDEA), 2-amino-2-methyl-1-propanol (AMP), etc. Among other small-molecule organic amines, ethylenediamine ($H_2NCH_2\ CH_2NH_2$) has the lowest price. With good injectability, it can selectively seal the channels of CO_2 gas channeling. Moreover, an organic gel sealing system is the main way to manage water and gas channeling. By adding cross-linking agents, the intermolecular cross-linking of polymers is established, forming a net-like structure of the gel system. Currently, several gels are used in combination, such as Cr^{3+} crosslinking, aluminum crosslinking, phenolic crosslinking. This type of gel system has poor applicability in reservoirs with high temperature and high salinity, low permeability, etc. The use of acrylamide copolymers with temperature-resistant and salt-resistant structural units and large molecules of multifunctional group cross-linkers to generate gels is in the research and development stage.

The most frequently studied and applied CO_2 mobility control is the foam system, which has good selective plugging ability for the formation with prominent inter- and intra-layer conflicts and large differences in oil saturation. It is characterized by low strength and short validity. At present, nanoparticle-stabilized and polymer-enhanced foam are hot spots of research concern. Some indoor and field studies on CO_2 foam injection methods have been carried out in different countries. For example, CO_2 foam field tests have been conducted in Rock Creek Oilfield, Virginia, and North Ward Estes Oilfield, Texas in the United States. In China, CO_2 foam flooding field tests have been conducted in Yumen, Karamay, Daqing, and Jidong oilfields. However, the stability of foam in the CO_2 foam injection method is affected by multiple factors such as pressure, temperature, residual oil saturation, and salts in formation water. There is no in-depth research on CO_2 foam flooding technology. The supporting technology of CO_2 foam injection is not ready in place, and the injection mechanism is not clear. This greatly increases the risk of CO_2 foam flooding.

It is difficult to have a long-lasting effect for gas channeling with the traditional plugging method. This is probably the main reason for not using traditional plugging in large-scale CO_2 flooding sealing applications. The chemical sealing technology for CO_2 flooding in China has just started. There are few reports of large-scale application in oilfields. As the research progresses, the future direction of CO_2 mobility control and sealing method is graded management and composite synergy, that is, to achieve effective sealing of CO_2 through the combination of multiple methods [38–40].

Gas channeling during the CO_2 EOR process cannot be avoided. However, by chemically assisted plugging, most of the injected CO_2 is retained in the subsurface, which can effectively expand the CO_2 sweep efficiency and improve the CO_2 EOR effect. In general, gas channeling requires hierarchical management (**Figure 6**). For high-permeability zones, a nano-modified stable foam system should be used for flow control, with a plugging rate of over 90%. For medium- and small-scale fractures, an organic-inorganic composite gel system can be used for chemical plugging, with a plugging rate of over 95%. For large-scale fractures, conventional methods are hardly effective. Adaptive gel particles can be used for bridging and plugging. By

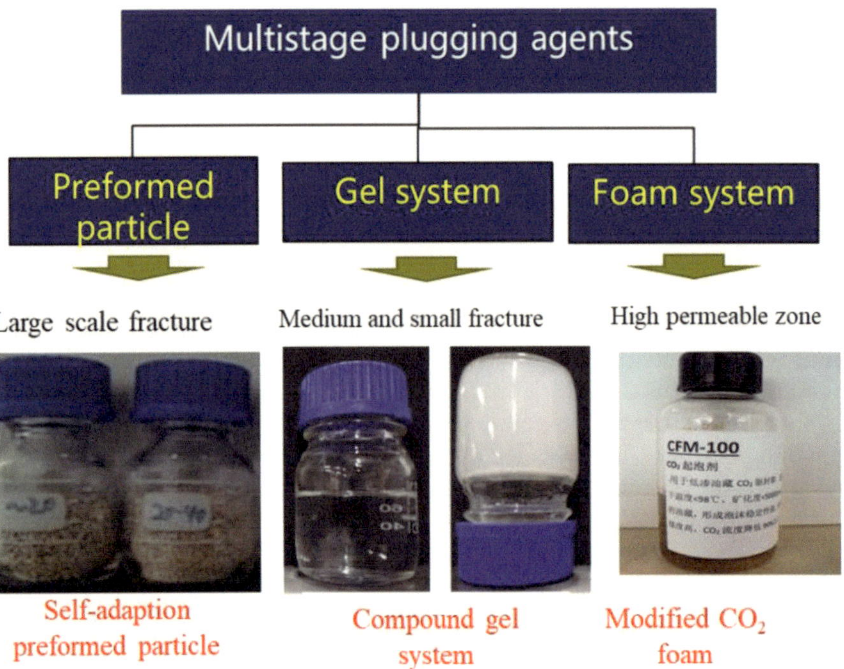

Figure 6.
CO_2 chemical enhancement system.

implementing hierarchical management through various methods, the extent of gas channeling can be effectively alleviated, and the gas displacement efficiency can be improved. At the same time, the CO_2 storage ratio can be greatly increased. Therefore, CO_2 channeling suppression is a technical guarantee for large-scale CO_2 storage in the process of oil displacement.

7. Full-chain optimized economic evaluation technology for CO_2 EOR and storage

7.1 Challenges for economic evaluation

As carbon capture, utilization, and storage (CCUS) technology is a cluster of many technologies, it is not only complex in terms of variety, technology classification, and cross-sectoral complexity, but also dynamic in terms of planning and development. Moreover, the economics of CCUS projects are significantly influenced by various factors such as international and domestic climate change policies and changes in international energy prices. These complex factors bring a great challenge for the large-scale deployment and application of this technology. In addition, it is difficult for the traditional deterministic economic evaluation methods to effectively plan for CCUS technologies under complex and uncertain conditions. For example, the different stages of CO_2 injection technology and the diversity of different injection methods can lead to uncertainty in CO_2 source requirements.

Traditional deterministic planning methods are difficult to respond and solve complex uncertainty problems effectively. With the continuous development of CCUS technology, a cross-regional storage utilization network from gas source to stored CO_2 will be formed. However, in the process of planning large-scale CCUS projects carried out across regions, the gas source supply at the gas gathering stations has a dynamic nature. Under the uncertain conditions of dynamic demand for different storage utilization methods (e.g., EOR utilization, saline formation storage utilization), CO_2 supply scheduling planning is required. Decision makers formulating the CO_2 allocation policy for each storage area based on the experiences will largely increase the risk of CO_2 oversupply or undersupply. In this complex uncertain environment, the CO_2 supply allocation scheduling planning is difficult to be managed effectively using traditional deterministic planning methods. Further coupling of uncertain optimization methods is required to develop effective models for rational planning of CCUS.

7.2 Different economic models

Scholars have proposed a variety of CCUS system optimization models to explore ways to reduce costs and investment decision risks. At present, the main economic models include the economic evaluation model for each CCUS link, the economic evaluation model for CCUS based on subsidies and incentives, the CCUS planning model based on uncertain optimization methods, and the planning model for CO_2 supply distribution scheduling.

7.2.1 Evaluation model for each CCUS link

The economic evaluation models for each link mainly focus on optimization analysis of CO_2 reduction cost, energy efficiency, and energy demand for pre-combustion, post-combustion, and oxygen-enriched combustion capture methods in coal-fired power plants. For the transportation and pipeline, the impact of different transportation methods on the cost and the impact of compression and energy consumption in the pipeline process are discussed. In terms of storage, the economic analysis is mainly focused on the CO2 EOR and storage processes. The integrated costs of produced CO_2 recovery, compression, reinjection, and oil displacement are mainly considered. A full-chain economic evaluation method for CO2 EOR and storage process is developed.

7.2.2 Evaluation model based on subsidies and incentives

The economic evaluation model of CCUS based on subsidies and incentives mainly considers incentives such as carbon taxes or fiscal subsidies to evaluate the economics of CCUS. Due to the high initial cost of CCUS technology, the funding problem cannot be solved by relying solely on corporate financing. External financing channels for CCUS technologies include market-based economic incentives and government financing incentives. As an important economic incentive for CCUS technology to reduce emissions, carbon trading has attracted much attention globally because of its flexible market mechanism. Domestic and international scholars have developed economic evaluation models for CCUS projects and analyzed the impact of carbon trading and incentive policies on CCUS investment decisions. In the past, many scholars have carried out economic evaluations for CCUS projects through external economic incentive policies, such as carbon trading prices. However, there is a lack of systematic studies considering the whole process of carbon capture, transportation, storage utilization, recycling, and reinjection. Consideration of the resource benefits and environmental benefits of large-scale GHG emission reduction resulting from the utilization of CCUS technology is missing as well. Therefore, it is difficult for the past studies to effectively evaluate CCUS technology and conduct CCUS full process planning studies with consideration of resource and environmental benefits [41–45].

7.2.3 CCUS planning model based on uncertain optimization methods

The CCUS planning model based on uncertainty optimization methods is designed to effectively solve the uncertainty problem in the CCUS planning and management process. Various uncertainty optimization methods have been developed, while the most widely used ones are three mathematical programming methods: interval programming, fuzzy programming, and stochastic programming. The interval programming method can effectively reflect the uncertainty information based on interval form in the engineering and technology planning process, which can provide a relatively stable solution for decision-makers. This method has been widely used in various fields such as energy planning, power, and atmosphere. The fuzzy mathematical programming method mainly introduces fuzzy sets into the mathematical programming method to deal with complex fuzzy uncertainty problems. This method is mainly applied to deal with fuzzy uncertainty problems of CCUS or to carry out the evaluation of a technical aspect of CCUS. Stochastic programming methods can effectively reflect the stochastic uncertainty of CCUS. The common stochastic programming methods include two-stage programming, multi-stage programming, chance-constrained programming, and stochastic robust programming. Few researchers have applied this method to the full process planning of CCUS. It was more frequently used to conduct planning research for a technical link of CCUS technology [46, 47].

7.2.4 Scheduling planning model for CO_2 supply distribution

The scheduling planning model based on CO_2 supply and allocation is mainly used to solve the issue of future network construction from gas source to storage utilization across regions. Researchers have carried out the development of CCUS transport optimization and scheduling planning models and achieved large-scale effective deployment of CCUS. For example, based on mixed-integer, multi-stage CO_2 transport and storage, a network model called the CCTS model was used to optimize the deployment of CO_2 capture, flow, and injection volumes. For saline aquifer storage and CO_2 flooding, a CO_2 pipeline network dynamic optimization allocation model is constructed and applied to optimize the whole project, mainly for selecting the most suitable CO_2 storage sites. In addition, there is also a pipeline network model based on source-sink matching optimization, which involves multiple sources and sinks. Optimal supply allocation is achieved from different emission sources such as power plants, cement plants, oil refineries, oilfields, coal beds, and saline aquifer storage.

7.3 Full-chain optimization model

Researchers have conducted a series of studies on CCUS technology from different perspectives. Their studies on strategic planning and techno-economic analysis of CCUS are relatively macro or one-sided, which can hardly guide the planning and techno-economic analysis at the project level. The development and application of CCUS-related models are more focused on a certain technology and process optimization. For example, the existing CCUS system planning models focus on two aspects: source-sink matching and pipe network optimization. The existing research lacks the lowest cost system optimization of the whole process project from the perspective of the system, according to the interaction between various technical links within the system. Particularly, the whole process system optimization research combined with CO_2 flooding oil utilization is missing. The current model is not a full-chain optimization model, that is, capture-compression-transport-injection-oil displacement (or gas displacement, coalbed methane, etc.)-recycle injection-long-term storage monitoring, etc. Particularly, the cost of cycle injection and monitoring are less considered. Therefore, it is difficult to reflect the lowest cost of the full-process CCUS-EOR project and the respective programming scheme based on the lowest cost. Due to the lack of support from relevant full-process system models, the past studies on the economics of CCUS based on carbon trading and government policy support can hardly provide a solid decision-making foundation. Some of the models do not evaluate the possible net regional emissions. As each link of CCUS consumes a certain amount of energy, which is equivalent to emitting a certain amount of CO_2, the net reduction of the whole region should be the amount of CO_2 stored minus the emissions of each link. Therefore, the future CCUS system planning model should be a full-chain optimization model based on net emission reduction and carbon footprint. In addition, the existing studies have failed to reflect the dynamic nature of CCUS project planning, uncertainty, and technical and commercial risks.

8. Conclusions

Generally, the development of CO_2 EOR technology mainly focused on two directions: improving the oil displacement effect and increasing the storage ratio. For improving the oil displacement effect, lots of studies were conducted on increasing the oil displacement efficiency and expanding the sweeping effect. By injecting water or gas in advance to replenish the formation energy, the formation pressure was raised above the minimum miscible pressure, thus achieving miscible flooding, and significantly increasing the oil displacement efficiency. In addition, the sweep efficiency of gas flooding can be effectively increased by improving the well pattern, matching different well types, and using appropriate reservoir modification measures. When injection and production parameters or process control measures fail to work, chemically assisted efficiency enhancement methods need to be considered. The addition of specific chemical agents can effectively reduce the minimum miscible pressure, block the gas channel, and increase the sweep efficiency, thus improving the gas flooding effect. To obtain an overall improvement of the field oil displacement effect, abundant indoor experimental studies are conducted. Based on the experimental results, corresponding numerical simulation methods are formed, and reasonable reservoir engineering and field implementation plans are designed. Generally, it is easier and less costly to implement by adjusting the injection and production parameters. On the contrary, it is less economical by injecting chemical agents, with high dosages of chemicals and high costs, probably leading to environmental problems as well, which conflicts with the environmental protection concept of CO_2 storage.

For improving the sequestration effect, the research is mainly focused on the scientific problem of how to increase the retention rate of CO_2 in the formation. This issue can be addressed by learning from the idea of flood management. First, taking the idea of "plugging," the gas channels are blocked in the production wells, allowing the gas to move to other parts of the formation, and thus achieving underground CO_2 retention. In addition, taking the idea of "unblocking," the produced gas is recovered and reinjected into the formation to achieve system-wide CO_2 retention. Currently, the latter idea is more commonly used in the field. That is, CO_2 is recycled by recovering from the produced gas and injecting it back into the formation. It can reduce the amount of newly purchased CO_2 and lower the cost of oil recovery. Meanwhile, the CO_2 retention rate is increased and the sequestration effect is improved as well.

Furthermore, the next generation of CO_2 EOR technology would focus on two aspects of the integrated CO_2 EOR and storage, the application of CO_2 EOR in unconventional resources. The CO_2 injection amount is more than 0.8–1.0 HCPV. The chemical-assisted method is more important. Additionally, the well pattern is more flexible. The storage coupling CO_2 EOR is required to be considered. The application of CO_2 EOR in tight oil, shale oil, and ROZ become the focus. The integrated CO_2 EOR and fracturing would be crucial in the development of these resources.

CO_2 EOR and storage are developing toward a direction that integrates enhanced oil recovery factor and CO_2 sequestration effect. As a complex process that incorporates gas injection, oil production, CO_2 storage, anti-corrosion technologies, planning, optimization, and implementation of CO_2 EOR and storage are challenging. The current experimental study of CO_2 EOR has been more concentrated on the physiochemical mechanisms, such as the interaction among CO_2, crude oil, brine, and rock. To be more realistic and flexible, the larger-scale experiments, such as microfluidic and physical simulation, are attracting more attention. For numerical simulation, the methods integrating the flow process with geochemical reaction and the variation of the stress field are preferred. Meanwhile, both CO_2 flooding and the sequestration mechanism should be equivalently considered in the simulation. The optimization design method has developed from the initial statistical method, through single objective optimization, to the current multi-objective optimization, which optimizes both oil displacement and

CO_2 storage. As key components of CO_2 EOR and storage technology, the supporting engineering technologies are discussed in detail, including gas injection, oil production, and anti-corrosion technologies. Particularly, the anti-corrosion technology is critical to the wellbore integrity and proper implementation of the CO_2 flooding process. Gas channeling is another main problem to be prevented during the implementation of CO_2 EOR and storage. Here, the gas channeling suppression technologies are summarized, including mechanical plugging, injection and production parameter adjustment, WAG injection, and chemical-assisted methods. With chemically assisted plugging, most of the injected CO_2 is retained in the subsurface, greatly increasing the CO_2 storage ratio. Therefore, CO_2 channeling suppression is a technical guarantee for large-scale CO_2 storage in the process of oil displacement. Although various studies on economic evaluation models for CCUS projects have been carried out in the past, it is challenging to integrally consider the whole process system optimization due to the inherent complexity and dynamic nature of planning and development of the CCUS process. Therefore, a full-chain optimization model based on net emission reduction and carbon footprint is suggested for the future system planning of CO_2 EOR and storage.

Acknowledgements

Thanks to the CCUS team of SINOPEC E&P Research Institute. Support for the National Key Research and Development program 2022YE0115800 and 2016YFB0600805-1. Support was provided by the National Science and Technology Major Project of the Ministry of Science and Technology of China 2016ZX05048003. The authors also show their deepest gratitude to Dr. Li Yang, the academician of China Engineering Academy.

Conflict of interest

The authors declare no conflict of interest.

Abbreviations

AMP	2-amino-2-methyl-1-propanol
CCUS	carbon capture, utilization and storage
CO_2	carbon dioxide
CT	computerized tomography
DEA	diethanolamine
EOR	enhanced oil recovery
GOR	gas oil ratio
HCPV	hydrocarbon pore volume
H_2NCH_2 CH_2NH_2	organic amines, ethylenediamine
MDEA	N-methyldiethanolamine
MEA	monoethanolamine
NMR	nuclear magnetic resonance
PVT	pressure-volume-temperature
TEA	triethanolamine
WAG	water alternative gas injection

Author details

Rui Wang*, Qiaoyu Ge, Zengmin Lun and Maolei Cui
SINOPEC Petroleum Exploration and Production Research Institute, Beijing, China

*Address all correspondence to: wangrui2009.syky@sinopec.com

References

[1] U.S. Department of Energy/National Energy Technology Laboratory, Improving Domestic Energy Security and Lowering CO_2 Emissions with "Next Generation" CO_2-Enhanced Oil Recovery (CO_2-EOR), DOE/NETL-2011/1504, prepared by Advanced Resources International, 2011. Available from: http://netl.doe.gov/energyanalyses/refshelf/PubDetails.aspx?Action=View&Source=Main&PubId=391

[2] Jarrel PM et al. Practical Aspects of CO_2 Flooding—Monograph Volume 22, Society of Petroleum Engineers, Richardson, Texas, 1st Printing; 2002. ISBN 1-55563-096-0

[3] Haynes S Jr, Alston RB. Study of the mechanisms of carbon dioxide flooding and applications to more efficient EOR projects. In: SPE/DOE Enhanced Oil Recovery Symposium. Tulsa, Oklahoma; 1990, SPE 20190-MS

[4] Martin DF, Tabor JJ. Carbon dioxide flooding. Journal of Petroleum Technology. 1992:396-400

[5] Koperna G, Wallace M, Reid T, et al. A Survey of U.S. CO2 Enhanced Oil Recovery Projects; 2021

[6] CO2 Enhanced Oil Recovery—Executive Summary. Institute for 21st Century Energy. USA; 2012

[7] Rui W, Zengmin L, Chengyuan L, et al. Research status and development trends of worldwide new technologies for enhanced oil recovery. Petroleum Geology and Recovery Efficiency. 2021;**28**(5):81-86

[8] Remson D. Storing CO2 and Producing Domestic Crude Oil with Next Generation CO2-EOR Technology[R]. National Energy Technology Laboratory; 2010

[9] Wallace M. The U.S. CO2 Enhanced Oil Recovery Survey[R]. Advanced Resources International Inc.; 2021

[10] Ghedan S. Global laboratory experience of CO2-EOR flooding. In: Paper SPE 125581 Prepared for Presentation at the 2009 SPE/EAGE Reservoir Characterization and Simulation Conference Held in Adu Dhabi, UAE. 2009

[11] Yellig WF, Metcalfe RS. Determination and prediction of CO2 minimum miscibility pressure. JPT. 1980:160-168

[12] Mansour EM, Al-Sabagh AM, Desouky SM, et al. A laboratory investigation of carbon dioxide-enhanced oil recovery by focusing on CO2-oil physical properties. Egyptian Journal of Petroleum. 2019;**28**:21-26

[13] Zekri AY et al. Experimental investigations of variations in petrophysical rock properties due to carbon dioxide flooding in oil heterogeneous low permeability carbonate reservoirs. Journal of Petroleum Exploration and Production Technology. 2013

[14] Bruce GH, Peaceman DW, Rachford HH, et al. Calculations of unsteady-state gas flow through porous media. Journal of Petroleum Technology. 1953;**5**(03):79-92

[15] Lee KS, Cho J, Lee JH. Numerical modeling of CO2 EOR. In: CO2 Storage Coupled with Enhanced Oil Recovery. Cham: Springer; 2020. DOI: 10.1007/978-3-030-41901-1_3

[16] Abass AE, Gawish AA, Elakhal EM. Simulation study of different modes of miscible carbon dioxide flooding. Egyptian Journal of Petroleum. 2018;**27**(4):1195-1207

[17] Yuan D, Wu Y, Binshan J, et al. A practical compositional method for simulation of CO2 flooding in porous and fractured petroleum reservoirs. In: Presentation at the SPE Asia Pacific Oil and Gas Conference and Exhibition Held in Jakarta, Indonesia. 2011

[18] Yeap WJ. Study of EOR-CO2 Miscible Flooding Performance Using Compositional Reservoir Simulation with Local Grid Refinement[D]. Thesis of Texas A&M University; 2018

[19] Qin T. EOS Modeling and Compositional Simulation Study of Carbon Dioxide Enhanced Oil Recovery in the Pembina Cardium Field, Alberta [Master's thesis], University of Calgary, Calgary, Canada; 2016. Available from: https://prism.ucalgary.ca. DOI: 10.11575/PRISM/25923

[20] Foroozesh J, Jamiolahmady M, Sohrabi M, et al. Non-equilibrium based compositional simulation of carbonated water injection EOR technique. In: 14th European Conference on the Mathematics of Oil Recovery. Catania, Sicily, Italy; 2014

[21] Rui W, Chengyuan L, Xuan L, et al. Experiments and simulation on integrated approach of CO2 EOR and storage in mature reservoirs. In: SPE Trinidad and Tobago Section Energy Resources Conference. 2021

[22] Wood D. Creating a Simplified Model Using Dimensionless Groups for CO2 EOR and Storage in Gulf Coast Reservoirs: Presented in Austin, Texas. GCCC Digital Publication Series #05-06; 2005

[23] Aoepov LE, Kaebov EN, Taboekov VG. Petrophysical Study of Oil Reservoir. Beijing: Petroleum Industry Press; 1992

[24] Chunsen Z, Bo T, Didi W. Water drive characteristic curve theory of low permeability reservoir. Advances in Petroleum Exploration and Development. 2015;**9**(1):83-85

[25] Hongen D, Hujun Z, Sibo S. Correct understanding and application of waterflooding characteristic curves. Petroleum Exploration and Development. 2019;**46**(4)

[26] Chengyuan L, Rui W, Shuxia Z, et al. Study on displacement characteristic curve in CO_2 immiscible flooding for low permeability reservoirs. Petroleum Geology and Recovery Efficiency. 2017;**24**(5):111-114

[27] Ampomah W, Balch R, Will R, et al. Co-optimization of CO2-EOR and storage processes under geological uncertainty. Energy Procedia. 2017;**114**:6928-6941

[28] Ettehadtavakkol A, Lake LW, Bryant SL. CO2-EOR and storage design optimization. International Journal of Greenhouse Gas Control. 2014;**25**:79-92

[29] Fleming EA, Brown LM, Cook RL. Overview of production engineering aspects of operating the Denver unit CO2 flood. In: SPE/DOE Symposium on Enhanced Oil Recovery. 1992, SPE-24157

[30] Bruckdorfer RA. Carbon dioxide corrosion in oilfield cements. In: SPE Rocky Mountain Regional Meeting. Billings, Montana; 1986, SPE-15176

[31] Linn LR. CO2 injection and production field facilities design evaluation and considerations. In: 62nd

Annual SPE Technical Conference and Exhibition. 1987, SPE-16830

[32] Bears DA, Wied RF, Martin AD, Doyle RP. Paradis CO2 flood gathering, injection and production systems. Journal of Petroleum Technology. 1984:1312-1323

[33] Gale J, Davidson J. Transmission of CO2—Safety and Economic Considerations. Sixth International Greenhouse Gas Technology Conference; 2003

[34] Meyer JP. Summary of Carbon Dioxide Enhanced Oil Recovery (CO2 EOR) Injection Well Technology. Prepared for the American Petroleum Institute; 2013

[35] Chauhan DS, Quraishi MA, Sorour AA, et al. A review on corrosion inhibitors for high-pressure supercritical CO2 environment: Challenges and opportunities. Journal of Petroleum Science and Engineering. 2022;**215**(Part B):110695

[36] Haofei S, Haoxiang W, Yimin Z, et al. Corrosion challenges in supercritical CO2 transportation, storage, and utilization—A review. Renewable and Sustainable Energy Reviews. 2023;**179**:113292

[37] Gan C, Ziqing Y, Jianguo L, et al. A comprehensive review of metal corrosion in a supercritical CO2 environment. International Journal of Greenhouse Gas Control. 2019;**90**:102814

[38] Ricky EX, Mwakipunda GC, Nyakilla EE, et al. A comprehensive review on CO2 thickeners for CO2 mobility control in enhanced oil recovery: Recent advances and future outlook. Journal of Industrial and Engineering Chemistry. 2023;**126**:69-91

[39] Massarweh O, Abushaikha AS. A review of recent developments in CO2 mobility control in enhanced oil recovery. Petroleum. 2022;**8**(3):291-317

[40] Wang R, Lv CY, Lun Z, et al. Study on gas channeling characteristics and suppression methods in CO2 flooding for low permeability reservoirs. In: Presented at the SPE Asia Pacific Oil and Gas Conference and Exhibition. Jakarta, Indonesia; 2011

[41] Zhao X, Xiao J, Hou J, et al. Economic and scale prediction of CO2 capture, utilization and storage technologies in China. Petroleum Exploration and Development. 2023;**50**(3):751-764

[42] Hong WY. A techno-economic review on carbon capture, utilisation and storage systems for achieving a net-zero CO2 emissions future. Carbon Capture Science & Technology. 2022;**3**:100044

[43] Morgan A, Ampomah W, Grigg R, et al. Techno-economic life cycle assessment of CO2-EOR operations towards net negative emissions at farnsworth field unit. Fuel. 2023;**342**:127897

[44] Ren B, Frank M, Duncan IJ. Economic analysis of CCUS: Accelerated development for CO2 EOR and storage in residual oil zones under the context of 45Q tax credit. Applied Energy. 2022

[45] Saha S, Saini G, Mishra S, et al. A comprehensive review of techno-socio-enviro-economic parameters, storage technologies, sizing methods and control management for integrated renewable energy system. Sustainable Energy Technologies and Assessments. 2022;**54**:102849

[46] Zheng Y, Gao L, Li S, et al. A comprehensive evaluation model for full-chain CCUS performance based on

the analytic hierarchy process method. Energy. 2022;**239**(Part D):122033

[47] Zhang F-Z, Xu R-N, He Y-F, et al. Analysis and optimization of energy flow in the full chain of carbon dioxide capture and oil recovery. Carbon Neutrality. 2022

Chapter 4

In-Situ Synthesis of Nanoparticles for Enhanced Oil Recovery (EOR) Operations: Current Status and Future Prospects

Reza Gharibshahi, Nafiseh Mehrooz and Arezou Jafari

Abstract

In-situ method synthesizes nanomaterials under reservoir conditions, harnessing the reservoir's energy. It offers several advantages over the alternative process of synthesizing these particles outside the reservoir and subsequently injecting them into the porous medium. This study provides an overview of the fundamentals, effective parameters, and mechanisms of this in-situ synthesis method. A comparison between in-situ and ex-situ synthesis of nanoparticles is presented, along with a discussion of their respective advantages and disadvantages. The impact of in-situ synthesis of nanoparticles on oil production and crude oil upgrading is thoroughly examined. It was observed that in-situ synthesis of nanoparticles leads to a uniform distribution of nanoparticles within the reservoir, thereby reducing issues related to formation damage. Furthermore, in-situ synthesized nanoparticles exhibit a superior ability to reduce the viscosity of crude oil, increase the API gravity, absorb asphaltenes, and enhance the oil recovery factor compared to the ex-situ synthesis method.

Keywords: nanoparticle, in-situ synthesis, EOR, viscosity, asphaltene, oil recovery factor

1. Introduction

Crude oil stands as a fundamental requirement for large-scale industries. Consequently, the optimal production and extraction of crude oil from hydrocarbon reservoirs represent pressing concerns within today's fuel supply industries [1]. In light of the limited oil resources and human capabilities for discovering new reservoirs, a significant challenge in the upstream sector of the oil industry is the production and exploitation of hydrocarbon resources in the latter stages of their lifespan, where production has dwindled [2]. This phase, known as Enhanced Oil Recovery (EOR), encompasses a broad spectrum of methods, including thermal techniques, chemical flooding, miscible and immiscible gas injection, microbial methods, and other contemporary approaches [3]. Each of these methods has its distinct advantages and disadvantages, along with operational limitations in field applications [4].

The emergence and extensive applications of nanotechnology in diverse industries such as electronics, biomedicine, pharmaceuticals, materials, aerospace, and more have prompted many experts in energy-related fields to invest in this technology [5]. Consequently, numerous researchers have turned to the use of nanoparticles to address various challenges within the oil industry, enhance the efficiency of conventional EOR methods, and increase production from oil reservoirs, particularly heavy and extra-heavy oil reservoirs [6, 7]. Nanotechnology deals with materials at dimensions exceedingly close to the molecular scale (nanometers, 10–9 m). It has offered novel solutions to longstanding problems that prior technologies struggled to resolve. As materials transition from bulk to nano-scale, their properties undergo substantial changes, governed by entirely distinct laws compared to micro and macro dimensions [8].

Within the nanometer scale, material properties become highly dependent on the size of their constituent components, resulting in nanoparticles exhibiting behaviors different from bulk materials [9]. By incorporating nanoparticles into a base fluid (such as water, oil, and gas) and forming a colloidal suspension of fine nanometer particles (known as nanofluid), several crucial properties of the base fluid, including thermal, hydrodynamic, magnetic, and intermediate stress properties, can be altered [10]. These changes are heavily contingent on the type, size, and shape of the nanoparticles.

The exceptional properties of nanomaterials can substantially enhance EOR performance [11]. Notably, nanomaterials possess high surface energy. By selecting appropriate nanoparticles in terms of type and size, it is possible to modify rock and fluid properties within the reservoir, facilitating easier oil production. In low-permeability reservoirs, capillary forces play a pivotal role in oil recovery within the minuscule pores and channels of the reservoir rock [12]. Conventional EOR methods focus on capillary, viscous, and gravitational forces, while nanotechnology relies on intermolecular and quantum forces [13].

Traditional chemical EOR methods often encounter challenges such as pore closure in porous media due to injected fluids (e.g., polymeric materials), formation damage, chemical wastage, and the degradation of material properties in deep reservoirs under harsh conditions [14]. Nanoparticles, owing to their minute size, can penetrate rock pores without causing formation damage or reducing the reservoir rock's permeability, thereby recovering oil droplets effectively. In essence, the unique properties of nanoparticles, such as their small size, high surface-to-volume ratio, robust mechanical and thermal resistance, excellent catalytic activity, high surface energy, and specific chemical and physical characteristics, have positioned nanoparticle injection into the reservoir as an efficient method for enhancing oil recovery [15]. Furthermore, nanoparticles are more environmentally compatible than other EOR methods and exhibit resistance to deformation under high temperatures and pressures, rendering them suitable for diverse reservoirs [16].

Despite extensive efforts, researchers face various challenges, including the industrial-scale and cost-effective production of nanomaterials, stability in high-salinity environments, and specialized equipment for injection, when developing the application of this method in large-scale EOR operations [17]. Hence, the development of optimal methods for the economical production of these materials in high quantities and a comprehensive understanding of their behavior in porous media can prove immensely beneficial [18].

Nanoparticles can be synthesized through various methods, each with its own set of advantages and disadvantages [19]. In recent years, a new approach has emerged for synthesizing nanoparticles by utilizing the reservoir's energy itself, known as the in-situ synthesis method [20]. It is anticipated that by refining this method and surmounting its challenges, the effectiveness of using nanoparticles in EOR processes can be significantly amplified. This endeavor may encourage oil companies to implement nanoparticles in actual oil fields. Consequently, this study aims to investigate the novel in-situ synthesis method of nanoparticles within the reservoir, elucidate its potential and mechanism, and review previous research in this area.

2. Mechanism of nanoparticles in EOR operation

A precise comprehension of the performance mechanisms of a substance plays a pivotal role in the success of an Enhanced Oil Recovery (EOR) plan [21]. Therefore, understanding how nanoparticles traverse porous media and influence oil production can significantly aid the advancement of this method in operational oil fields [22]. Various parameters, including the type, size, and concentration of nanoparticles, reservoir temperature, and pressure, base fluid salinity, shear rate, injection rate, reservoir rock properties, rock pore diameter, crude oil composition, asphaltene content, water content, and the duration of contact between asphaltene and nanoparticles, impact the performance of nanoparticles in an EOR process [23]. Research conducted thus far indicates that nanoparticles enhance oil recovery through various mechanisms [24–26]. Some of these mechanisms are elaborated below:

- Disjoining Pressure: Nanoparticles create a wedge-shaped layer at the three-phase water/oil/rock contact surface. This generates an osmotic pressure from the water phase toward the oil phase, prompting the movement of oil out of the rock pores.

- Pore Channels Plugging: Nanoparticles accumulate at the openings of smaller pores, temporarily blocking them. This intensifies fluid flow to other pores and facilitates oil recovery from them.

- Reducing Fluid Mobility: The addition of nanoparticles to the base fluid elevates its viscosity and reduces its mobility ratio. Consequently, injected fluid moves more effectively within the reservoir, lowering the likelihood of fingering effects.

- Preventing and Controlling Asphaltene Deposition: Nanoparticles adsorb asphaltene molecules on their surfaces, enhancing asphaltene stability in the oil phase and diminishing their precipitation and separation from the oil phase.

- Lowering Interfacial Surface Tension (IFT): Nanoparticles placed at the water–oil interface reduce capillary pressure between them, thereby reducing IFT between the two phases. This enhances the distribution and movement of fluids within the porous medium.

- Wettability Alteration: Nanoparticles can alter the reservoir rock surface from oil-wet to strongly water-wet, overcoming capillary forces and facilitating the easier separation of oil droplets from the rock surface.

- Stability of Foam/Microemulsions: Nanoparticles covering the interface between two phases (e.g., water/gas or water/oil) within the reservoir enhance surface elasticity, improving the stability of foam and emulsions.

- Improved Thermal Properties: The addition of metal nanoparticles to the base fluid enhances thermal properties such as thermal conductivity and heat capacity. This leads to improved heat distribution within the reservoir and increased efficiency of thermal EOR methods.

- Upgrading Oil Quality: Due to their unique catalytic properties, nanoparticles absorb large molecules in crude oil, such as asphaltene and resin, on their surfaces. Through catalytic cracking, they reduce crude oil viscosity and enhance its quality, making it easier for the oil to flow toward the production well.

3. Synthesis of nanoparticles

In all methods involving the injection of nanofluids to enhance oil recovery, nanoparticles are initially synthesized on the reservoir's surface or outside of it [27]. Subsequently, these nanoparticles are introduced into the reservoir through the preparation of a colloidal solution (by dispersing them in a suitable base fluid and creating a stable nanofluid) [28]. These methods are known as ex-situ synthesis of nanoparticles. There exist various techniques for producing nanoparticles, which can be broadly categorized into the following three groups:

3.1 Physical vapor deposition

In this process, a solid metal is vaporized and then rapidly condensed to form nanoscale clusters and particles, which ultimately settle as powder. The most significant advantage of the Physical vapor deposition method is its low environmental impact. Moreover, this method allows for precise control of particle size by adjusting parameters such as temperature, gas medium, evaporation, and condensation rates, enabling the production of smaller and more uniformly sized particles [29].

3.2 Chemical methods

This category of synthesis methods involves two concurrent processes: nucleation and crystal growth in a liquid medium containing various reagents. By carefully regulating these processes, nanoparticles can be synthesized with a uniform size distribution and optimal size. This category encompasses various techniques such as sol-gel, co-precipitation, hydrothermal, solvothermal, sonochemical, microemulsion, microwave-assisted synthesis, and more. These methods are suitable for producing nanoparticles in high quantities and volumes at a relatively low cost. However, potential chemical pollution is a concern associated with these methods [30, 31].

3.3 Solid state processes

These methods entail the production of nanoparticles from larger-sized materials through processes like grinding or powdering. Various parameters, such as the type of grinding material, grinding time, and atmospheric conditions, can influence the properties of the resulting nanoparticles. This method can be employed to produce nanoparticles that may be challenging to obtain using the previously mentioned methods. It is worth noting that this approach tends to have higher costs and lower production volumes of nanoparticles compared to the methods in the previous two categories [32, 33].

4. In-situ synthesized of nanoparticles for EOR application

Ex-situ preparation and injecting of nanofluids into the reservoir can lead to formation damage and a reduction in reservoir permeability due to the agglomeration of nanoparticles upon contact with formation water and their sedimentation in areas near the injection well [34]. Consequently, it is crucial to modify the surface of nanoparticles to enhance their stability in the base fluid. The presence of solid nanoparticles in the base fluid necessitates specialized equipment for nanofluid injection into the reservoir, raising concerns about equipment corrosion. Additionally, discussions regarding the cost of synthesizing these particles, the reinjection process into the reservoir, and potential health hazards during this transfer have posed challenges associated with the use of ex-situ nanoparticle synthesis methods in EOR processes [35].

In recent years, researchers have turned to in-situ synthesis methods to address the challenges associated with ex-situ methods for nanoparticle synthesis. The goal of in-situ nanoparticle synthesis is to employ precursor salts and introduce them into the reservoir. In this method, the reservoir's heat serves as the energy source for conducting the synthesis reaction.

It appears that various factors, such as the type and concentration of precursor salts, reservoir temperature and pressure, the type of reservoir rock and its surface charge, and the presence of specific ions, can influence the in-situ synthesis of nanoparticles. It's noteworthy that this synthesis method is a one-step process, as the introduction of additional substances into the reaction system becomes unfeasible once fluids are injected into the reservoir. In this method, precursor salts are initially dissolved in the base fluid in specific quantities, tailored to the type of nanoparticle. This fluid is then injected into the reservoir, leading to the formation of a water-in-oil microemulsion. Consequently, the most commonly employed method for in-situ nanoparticle synthesis is the microemulsion method (**Figure 1**), known for its effectiveness in controlled nanoparticle synthesis [37].

In this method, two immiscible solvents are employed in the presence of a surfactant for nanoparticle synthesis. The addition of a surfactant to the base fluid significantly reduces the IFT between oil and water, resulting in the formation of a stable microemulsion [38]. Crude oil may contain natural surfactants, such as resin and asphaltene, which can fulfill this role in an oil reservoir. The microemulsion method is suitable for synthesizing nanoparticles with uniform size, morphology, excellent dispersibility, and various shapes. Particle size can be controlled by adjusting the oil/water ratio, temperature, and the aqueous phase, while particle morphology can be influenced by altering the pH of the reaction mixture [39].

Subsequently, the aqueous phase droplets move within the emulsion, merging and mixing. The synthesis reaction commences as reactive molecules penetrate from one droplet to another. The energy required for nanoparticle synthesis is derived from the heat within the reservoir. As the primary nucleus forms and grows, the synthesis reaction concludes, yielding nanoparticles. The droplets are then separated from one another once more. Consequently, this method is proficient at producing monodispersed nanoparticles characterized by small, uniform particle size.

5. In-situ vs. ex-situ synthesis of nanoparticles and effective mechanisms

In-situ nanoparticle synthesis, due to the smaller size and heightened surface activity of nanoparticles, can be more effective in crude oil production mechanisms

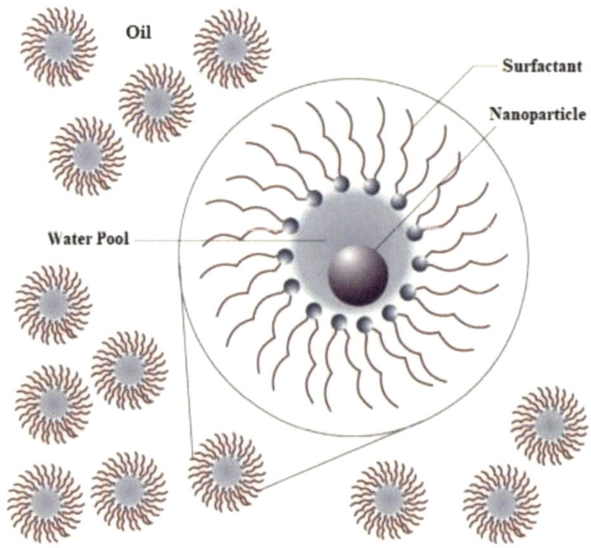

Figure 1.
Schematic representation of water-in-oil microemulsion [36].

from porous media, such as reducing crude oil viscosity. This method employs precursor salts for nanoparticles, allowing the use of conventional equipment and pumps for their injection into the reservoir. Nanoparticles synthesized using this method exhibit greater stability, resolving issues related to nanoparticle deposition within reservoir rock pores, permeability reduction, and formation damage to a significant extent. This is because nanoparticles are initially synthesized through ex-situ processes. Subsequently, to disperse them in the base fluid, typically water due to its abundance, the surface of these nanoparticles must undergo hydrophilic modification to facilitate the preparation of a stable colloidal fluid. However, practical experience has demonstrated that the application of mechanical stresses within the injection equipment, coupled with the high pressure and temperature conditions within the reservoir, as well as the salinity of the reservoir fluids, leads to a rapid and substantial reduction in the stability of these nanofluids. Consequently, solid particles become trapped within the pores of the reservoir rock, significantly diminishing permeability due to particle deposition within the throat regions of the reservoir rock. This issue gives rise to problems associated with formation damage, which necessitates consideration by experts in a protective production process.

However, in in-situ processes, nanoparticles are produced utilizing their precursor salts. These particles are synthesized under reservoir conditions, at the reservoir's temperature and pressure, in tandem with the movement of the injected fluid within the porous medium. The smaller size of the particles in this method, along with the resulting high surface-to-volume ratio and surface energy, facilitates their easy penetration into the small pores of the reservoir rock, allowing for efficient oil removal without entrapment. Consequently, in the in-situ synthesis processes of nanoparticles, the risk of particle sedimentation within the reservoir rock pores and subsequent permeability reduction is significantly reduced.

Additionally, nanoparticles are more uniformly distributed in the reservoir fluids, increasing the stimulated area within the reservoir and enabling nanofluids to affect a larger volume of the porous medium. This is because in the in-situ synthesis processes of nanoparticles, precursor salts of the desired material are utilized. These salts exhibit complete solubility in water, significantly reducing the likelihood of precipitation within the pores of the reservoir rock when compared to ex-situ synthesized nanoparticles. Since nanoparticles are gradually synthesized in the in-situ method, harnessing the reservoir's energy during the injection process, the injection fluid can uniformly transport a larger quantity of nanoparticles into a greater reservoir volume. Consequently, in flooding processes, a more extensive portion of the reservoir will be stimulated to enhance crude oil production. Furthermore, this approach streamlines the synthesis of nanoparticles, utilizing the temperature and energy available within the reservoir itself for synthesis reactions. Consequently, costs associated with nanoparticle synthesis, reinjection, and economic concerns, including material wastage, nanoparticle transportation, and potential human health risks, are substantially reduced.

These attributes underscore the importance of studying and understanding the mechanisms of in-situ synthesized nanoparticles, surpassing the significance of the ex-situ synthesis method. If the challenges associated with optimal in-situ nanoparticle synthesis can be successfully addressed, many issues in field-scale nanofluid injection operations can be overcome. **Figure 2** summarizes the advantages of this method.

6. Effect of in-situ synthesis of nanoparticles on oil production and upgrading

In recent times, a select group of researchers has delved into the possibility of in-situ nanoparticle synthesis within the oil reservoir, operating independently of the reservoir rock. Studies conducted in this realm have revealed that a limited set of nanoparticles, namely iron oxide (Fe_2O_3), nickel oxide (NiO), vanadium oxide (V_2O_5), alumina (Al_2O_3), copper oxide (CuO), and cerium oxide (CeO_2), have been successfully synthesized in-situ to enhance oil recovery. The prevailing method employed for synthesis primarily involves the creation of a stable water-in-oil microemulsion from precursor salts. Typically, researchers have utilized Parr reactors or high-pressure reactors, essential not only for maintaining the high temperature

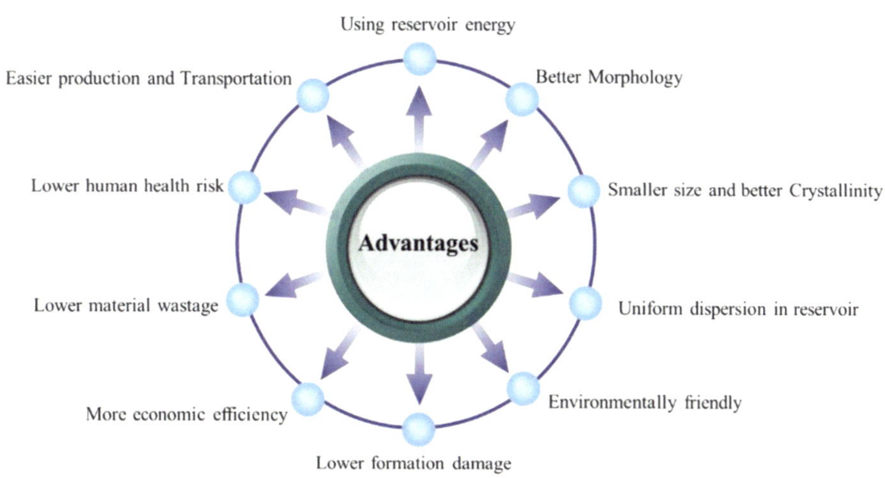

Figure 2.
Advantages of the in-situ synthesis of nanoparticles.

required but also for achieving the necessary pressure for the synthesis reaction. The synthesized nanoparticles have fallen within the size range of 5 to 35 nm. Furthermore, various objectives, such as asphaltene absorption, viscosity reduction, and enhancement of the oil recovery factor, have been thoroughly investigated. The following table provides an overview of the studies conducted in the field of in-situ nanoparticle synthesis for the purpose of improving oil production.

Previous studies have demonstrated that the control of crystallinity and morphology of nanoparticles using the in-situ synthesis method surpasses that of the ex-situ method. Abdrabo et al. [40] achieved in-situ synthesis of NiO nanoparticles with a size of 20 nm and V_2O_5 nanoparticles with a size of 15 nm using ammonium metavanadate (NH_4VO_3) and nickel (II) nitrate ($Ni(NO_3)_2$) precursor solutions. By carefully regulating the size of synthesized nanoparticles, the active surface area of these nanoparticles increases significantly. This enhanced catalytic activity enables the breakdown of heavy crude oil molecules and the upgrading of heavy crude oil and bitumen. Additionally, Biyouki et al. [41] illustrated that in-situ synthesized NiO nanoparticles exhibit superior crystallinity and morphology compared to commercial nanoparticles. They revealed that nucleation and growth mechanisms for NiO crystals occur in the liquid phase, leading to condensation, primary nucleus accumulation, secondary particle formation, and nanoparticle crystal growth.

Mehrooz et al. [20] adopted a straightforward one-step approach within a crude oil medium for the in-situ synthesis of CeO_2 nanoparticles at low temperatures. They employed precursor salts such as ammonium nitrate (NH_4NO_3) and Ceric ammonium nitrate ($H_8N_8CeO_{18}$) to create a stable water-in-oil microemulsion (see **Figure 3**). They systematically investigated the impact of various parameters, including temperature, pH, precursor salt concentration, and stirring time, on the size and quality of the in-situ synthesized CeO_2 nanoparticles. Their method successfully generated high-quality CeO_2 nanoparticles with an average size of 18 nm. Their findings underscored that the reaction temperature of the solution exerted the most significant influence on the size of the synthesized nanoparticles.

Husein et al. [43] employed an in-situ method to synthesize Al_2O_3 nanoparticles with an average diameter of 17 nm at a temperature of 300°C. Thanks to their higher

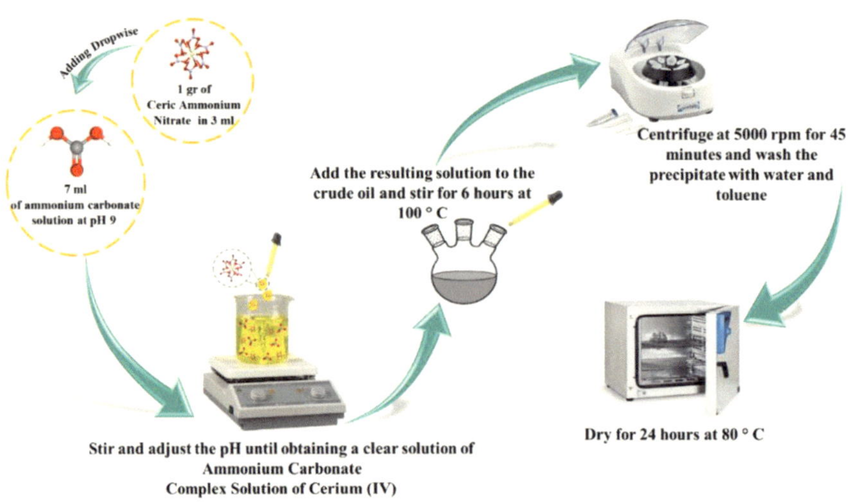

Figure 3.
Proposed procedure for in-situ synthesis of CeO_2 nanoparticles [42].

dispersion surface and more uniform distribution, these in-situ synthesized Al_2O_3 nanoparticles exhibited superior catalytic activity in reducing viscosity and increasing the API of crude oil compared to commercial nanoparticles. Hashemi et al. [44] employed trimetallic nanocatalysts of tungsten, nickel, and molybdenum with excellent dispersibility in vacuum gas oil (VGO). Their findings indicated that these nanoparticles possessed favorable catalytic properties for steam injection processes. In-situ synthesized nanoparticles increased the API of oil more effectively than commercial counterparts, ultimately increasing the production of heavy oil and bitumen by reducing viscosity. Chen et al. [45] in-situ converted copper hydroxide ($Cu(OH)_2$) precursor salt into CuO nanoparticles with high dispersion. CuO nanoparticles reduced the viscosity of heavy crude oil by 94.6% and concurrently converted 22.4% of asphaltenes into lighter components. Notably, minimal agglomeration of CuO nanoparticles was observed during catalytic reactions.

In comparison to commercial nanoparticles, in-situ synthesized nanoparticles exhibit a higher capacity to absorb asphaltene on their surfaces. Abu Tarboush and Husein [46] observed that in-situ synthesized NiO nanoparticles absorbed a larger amount of asphaltene on their surface than their commercial counterparts. In the case of in-situ synthesized nanoparticles, the amount of asphaltene absorption on their surface reached 2.85 (grams of asphaltene per gram of nanoparticle). In contrast, commercial nanoparticles could only absorb 15% of this amount. They also introduced a microemulsion method for in-situ synthesis of Fe_2O_3 nanoparticles. Their research indicated that the water content in the microemulsion had no significant

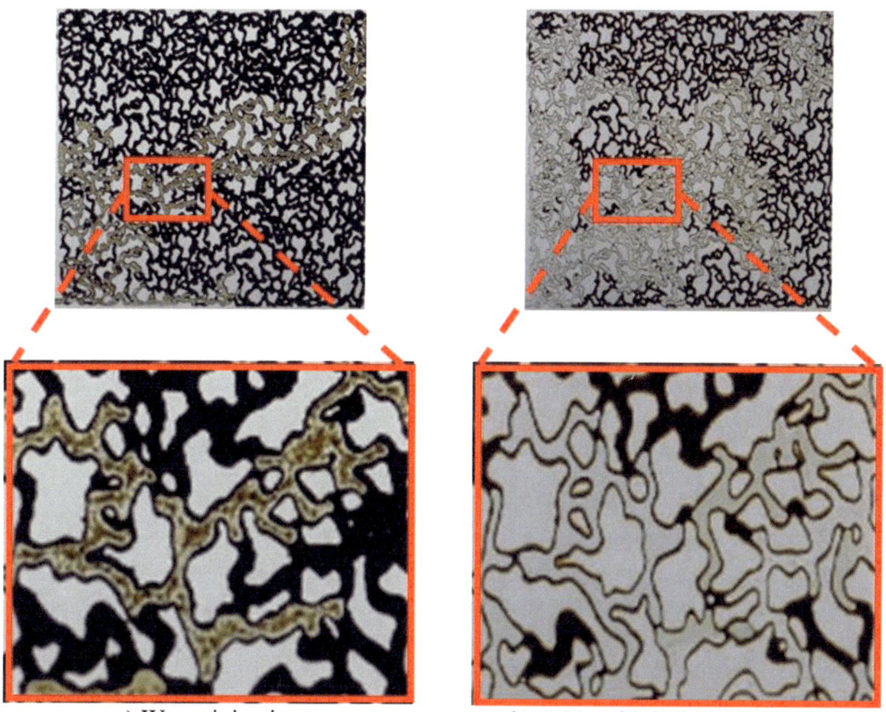

a) Water injection　　b) 0.5 wt. % CeO_2 nanoparticles

Figure 4.
The oil trapping effect during in-situ synhesiezed CeO_2 nanoparticles [42]. (a) Water injection (b) 0.5 wt. % CeO_2 nanoparticles.

impact on asphaltene absorption. Furthermore, they noted that asphaltene absorption on the surface of in-situ synthesized nanoparticles was highly favorable [47]. Biyouki et al. [41] found that in-situ synthesized NiO nanoparticles outperformed ex-situ synthesized NiO nanoparticles in the process of coke oxidation and preventing asphaltene precipitation.

In-situ nanoparticle synthesis can significantly enhance oil production, increasing the tertiary oil recovery factor from 10 to 28.5% using in-situ synthesized Fe_2O_3 nanoparticles at a concentration of 6400 ppm [48]. By introducing in-situ synthesized CeO_2 nanoparticles into the water as a dispersion medium, the oil recovery factor improved substantially. Additionally, increasing the concentration of CeO_2 nanoparticles in water reduced the fingering phenomenon. It is essential to note that exceeding the optimal nanoparticle concentration in the base fluid increases the likelihood of particle deposition within the porous medium. Simultaneously, the impact of nanoparticles on altering the hydrodynamic properties of the injected fluid and the mechanisms for improving oil recovery diminishes. Furthermore, CeO_2 nanoparticles exhibited an excellent ability to separate oil droplets adhered to the porous medium's walls, reducing the amount of crude oil trapping during the flooding process. This phenomenon was clearly observed in microscopic images of the porous medium (see **Figure 4**).

7. Challenges and future prospects

In spite of the extensive efforts and research conducted, the field of in-situ nanoparticle synthesis is still in its infancy, with only a limited number of studies focused on increasing oil production. Many aspects of this process remain incompletely defined, demanding closer attention from researchers. The synthesis of nanoparticles with specific and desirable characteristics, along with an in-depth investigation of the parameters influencing the optimal and high-quality synthesis of these particles for enhanced oil recovery (EOR) applications, should be a primary focus. Only through such efforts can this method be effectively employed at a field-scale EOR level. Several of the challenges and opportunities in this domain are as follows:

- As observed, only a limited range of nanoparticles has been synthesized using this method. It appears that inorganic nanoparticles, such as silica or clay, exhibit strong compatibility with oil reservoirs and possess significant potential for improving oil recovery factors. Consequently, the in-situ synthesis of various other types of nanoparticles and an examination of their performance should be pursued.

- The predominant synthesis method employed in this field has been the creation of a stable water-in-oil microemulsion. The exploration of alternative techniques, such as microwave and ultrasound waves, could prove beneficial. Consequently, there is a need for the development of a straightforward and cost-effective method for in-situ synthesis across diverse nanoparticle types.

- Considering that the majority of oil reservoirs maintain temperatures below 120°C, and this method relies on harnessing reservoir energy, researchers should emphasize nanoparticle synthesis at lower temperatures.

- A detailed exploration of the impact of various parameters on the in-situ synthesis of nanoparticles, including temperature, pressure, pH, type and concentration of precursor, stirring time, etc., is lacking.

- The precise effects of in-situ synthesized nanoparticles on asphaltene absorption and deposition in oil reservoirs, reduction of IFT, alteration of reservoir rock wettability, and crude oil viscosity have not been comprehensively examined.

- The influence of nanoparticle shape and morphology (e.g., cubic or spherical) on oil recovery factors remains unexplored.

- One of the key challenges in utilizing nanoparticles for EOR is ensuring their colloidal stability under extreme temperature and pressure conditions. Additionally, the salinity of the injection fluid can impact nanoparticle effectiveness. Therefore, it is imperative to develop suitable methods for enhancing the stability of in-situ synthesized nanoparticles. Furthermore, thorough investigation into the impact of salt type and concentration dissolved in the injection fluid on nanoparticle stability is warranted.

- For a more realistic assessment of the process, researchers should employ porous media such as sandpacks and core samples to conduct studies closely mimicking actual oil reservoir conditions, including the effect of reservoir rock on in-situ nanoparticle synthesis.

- Comprehensive studies on the environmental implications, subsurface water quality, and potential human health effects resulting from this method have yet to be conducted.

- One of the primary challenges in implementing nanoparticles in EOR processes is their costliness. Therefore, an economic feasibility study and strategies to reduce injection operation costs are crucial.

- Molecular simulation represents a viable approach for gaining insights into the mechanisms governing in-situ synthesized nanoparticles at microscopic scales.

8. Conclusions

In-situ synthesis of nanoparticles represents a novel approach to the economical production of nanoparticles and overcoming the challenges associated with ex-situ nanoparticle injection in Enhanced Oil Recovery (EOR) processes. In this method, the synthesis reaction takes place within the oil medium, leveraging the reservoir's energy. The principal advantages of this approach include the even distribution of nanoparticles within the reservoir, minimized formation damage, and greater impact on crude oil production mechanisms. When compared to the ex-situ method, the in-situ method affords more precise control over the crystallinity and morphology of nanoparticles. Size control assumes a pivotal role in determining the catalytic activity of in-situ synthesized nanoparticles. In-situ synthesized nanoparticles exhibit enhanced capabilities in reducing crude oil viscosity, increasing crude oil API, absorbing asphaltene molecules, and augmenting oil recovery factors in comparison to their commercial counterparts. Nevertheless, researchers face various challenges and opportunities in their endeavors to implement this method in real-world oil fields.

Author details

Reza Gharibshahi, Nafiseh Mehrooz* and Arezou Jafari
Faculty of Chemical Engineering, Tarbiat Modares University, Tehran, Iran

*Address all correspondence to: nafiseh.mehrooz@modares.ac.ir

References

[1] Kazemzadeh Y, Shojaei S, Riazi M, Sharifi M. Review on application of nanoparticles for EOR purposes: A critical review of the opportunities and challenges. Chinese Journal of Chemical Engineering. 2019;**27**:237-246. DOI: 10.1016/j.cjche.2018.05.022

[2] Gharibshahi R, Jafari A, Haghtalab A, Karambeigi MS. Application of CFD to evaluate pore morphology effect in the Nanofluid flooding for enhanced oil recovery. RSC Advances. 2015;**5**:28938-28949. DOI: 10.1039/C4RA15452E

[3] Gharibshahi R, Omidkhah M, Jafari A, Fakhroueian Z. Hybridization of superparamagnetic Fe3O4 nanoparticles with MWCNTs and effect of surface modification on electromagnetic heating process efficiency: A microfluidics enhanced oil recovery study. Fuel. 2020;**282**:118603. DOI: 10.1016/j.fuel.2020.118603

[4] Yakasai F, Jaafar MZ, Bandyopadhyay S, Agi A. Current developments and future outlook in nanofluid flooding: A comprehensive review of various parameters influencing oil recovery mechanisms. Journal of Industrial and Engineering Chemistry. 2021;**93**:138-162. DOI: 10.1016/j.jiec.2020.10.017

[5] Liang T, Hou JR, Qu M, Xi JX, Raj I. Application of nanomaterial for enhanced oil recovery. Petroleum Science. 2021;**19**(2):882-899. DOI: 10.1016/j.petsci.2021.11.011

[6] Tohidi Z, Teimouri A, Jafari A, Gharibshahi R, Omidkhah MR. Application of Janus nanoparticles in enhanced oil recovery processes: Current status and future opportunities. Journal of Petroleum Science and Engineering. 2022;**208**:109602. DOI: 10.1016/j.petrol.2021.109602

[7] Al-Asadi A, Rodil E, Soto A. Nanoparticles in chemical EOR: A review on flooding tests. Nanomaterials. 2022;**12**(23):4142. DOI: 10.3390/nano12234142

[8] Di Ventra M, Evoy S, Heflin JR. Introduction to Nanoscale Science and Technology. Vol. 6. Boston, United States: Springer Science & Business Media; 2004. pp. 1-3. DOI: 10.1007/1-4020-7757-2_1

[9] Panchal H, Patel H, Patel J, Shah M. A systematic review on nanotechnology in enhanced oil recovery. Petroleum Research. 2021;**6**(3):204-212. DOI: 10.1016/j.ptlrs.2021.03.003

[10] Gharibshahi R, Jafari A, Omidkhah M, Nezhad JR. Performance experimental investigation of novel multifunctional nanohybrids on enhanced oil recovery. AIP Conference Proceedings. 2018;**1920**(1):20012. DOI: 10.1063/1.5018944

[11] Khoramian R, Kharrat R, Golshokooh S. The development of novel nanofluid for enhanced oil recovery application. Fuel. 2022;**311**:122558. DOI: 10.1016/j.fuel.2021.122558

[12] Cheraghian G, Rostami S, Afrand M. Nanotechnology in enhanced oil recovery. PRO. 2020;**8**(9):1073. DOI: 10.3390/pr8091073

[13] Farshadfar H, Armandi HS, Gharibshahi R, Jafari A. Simultaneous electromagnetic radiation and Nanofluid injection and their interactions in EOR operations: A comprehensive review. Journal of Magnetism and Magnetic

Materials. 2023;**580**(1):170863. DOI: 10.1016/j.jmmm.2023.170863

[14] Gharibshahi R, Jafari A. Numerical investigation of controllable parameters effect on Nanofluid flooding in a random pore generated porous medium. Iranian Journal of Chemistry and Chemical Engineering. 2021;**40**(3):780-795. DOI: 10.30492/IJCCE.2020.66919.2560

[15] Gharibshahi R, Omidkhah M, Jafari A, Mehrooz N. Parametric optimization of in-situ heavy oil upgrading using simultaneous microwave radiation and magnetic nanohybrids via Taguchi approach. Fuel. 2022;**325**:124717. DOI: 10.1016/j.fuel.2022.124717

[16] Ayatollahi S, Zerafat M. Nanotechnology-assisted EOR techniques: New solutions to old challenges. In: SPE International Oilfield Nanotechnology Conference and Exhibition. June 2012; Noordwijk, Netherlands. 2012. pp. 1-15

[17] Fakoya MF, Shah SN. Emergence of nanotechnology in the oil and gas industry: Emphasis on the application of silica nanoparticles. Petroleum. 2017;**3**(4):391-405. DOI: 10.1016/j.petlm.2017.03.001

[18] Peng B, Tang J, Luo J, Wang P, Ding B, Tam KC. Applications of nanotechnology in oil and gas industry: Progress and perspective. Canadian Journal of Chemical Engineering. 2018;**96**(1):91-100. DOI: 10.1002/cjce.23042

[19] El-Khawaga AM, Zidan A, Abd El-Mageed AIA. Preparation methods of different nanomaterials for various potential applications: A review. Journal of Molecular Structure. 2023;**1281**:135148. DOI: 10.1016/j.molstruc.2023.135148

[20] Mehrooz N, Gharibshahi R, Jafari A, Delavari H, Sadeghnejad S. Low-temperature in-situ synthesis of cerium oxide nanoparticles applicable in EOR processes: Parametric study by Taguchi approach. Journal of Petroleum Science and Engineering. 2022;**219**:111078. DOI: 10.1016/j.petrol.2022.111078

[21] Sheng JJ. Enhanced Oil Recovery Field Case Studies. 1st ed. Amsterdam, Netherlands: Gulf Professional Publishing; 2013. p. 712. DOI: 10.1016/C2010-0-67974-0

[22] Boul PJ, Ajayan PM. Nanotechnology research and development in upstream oil and gas. Energy Technology. 2020;**8**(1):1901216. DOI: 10.1002/ente.201901216

[23] Sun X, Zhang Y, Chen G, Gai Z. Application of nanoparticles in enhanced oil recovery: A critical review of recent Progress. Energies. 2017;**10**(3):345. DOI: 10.3390/en10030345

[24] Shamsi Armandi H, Jafari A, Gharibshahi R. Nanoparticles assisted microwave radiation: Fluid-rock interactions in oil reservoirs. Petroleum Science. 2021;**18**(6):1797-1812. DOI: 10.1016/j.petsci.2021.09.002

[25] Saha R, Tiwari P, Uppaluri RVS. Chemical Nanofluids in Enhanced Oil Recovery: Fundamentals and Applications. Boca Raton, Florida, USA: CRC Press; 2021. p. 170. DOI: 10.1201/9781003010937

[26] Iravani M, Khalilnezhad Z, Khalilnezhad A. A review on application of nanoparticles for EOR purposes: History and current challenges. Journal of Petroleum Exploration and Production Technology. 2023;**13**(4):959-994. DOI: 10.1007/s13202-022-01606-x

[27] Gharibshahi R, Jafari A, Asadzadeh N. Influence of different

nanoparticles on the gas injection performance in EOR operation: Parametric and CFD simulation study. Canadian Journal of Chemical Engineering. 2023;**1**:1-14. DOI: 10.1002/cjce.25071

[28] Sun Y, Yang D, Shi L, Wu H, Cao Y, He Y, et al. Properties of nanofluids and their applications in enhanced oil recovery: A comprehensive review. Energy & Fuels. 2020;**34**(2):1202-1218. DOI: 10.1021/acs.energyfuels.9b03501

[29] Wegner K, Walker B, Tsantilis S, Pratsinis SE. Design of metal nanoparticle synthesis by vapor flow condensation. Chemical Engineering Science. 2002;**57**(10):1753-1762. DOI: 10.1016/S0009-2509(02)00064-7

[30] Baig N, Kammakakam I, Falath W. Nanomaterials: A review of synthesis methods, properties, recent progress, and challenges. Materials Advances. 2021;**2**(6):1821-1871. DOI: 10.1039/D0MA00807A

[31] Khan I, Saeed K, Khan I. Nanoparticles: Properties, applications and toxicities. Arabian Journal of Chemistry. 2019;**12**(7):908-931. DOI: 10.1016/j.arabjc.2017.05.011

[32] Aswani R, Radhakrishnan EK. In: Shanker U, Hussain CM, Rani M, editors. Green Approaches for Nanotechnology. Green Functionalized Nanomaterials for Environmental Applications. Amsterdam, Netherlands: Elsevier; 2022. pp. 129-154. DOI: 10.1016/C2019-0-04822-4

[33] Joy J, Krishnamoorthy A, Tanna A, Kamathe V, Nagar R, Srinivasan S. Recent developments on the synthesis of nanocomposite materials via ball milling approach for energy storage applications. Applied Sciences. 2022;**12**(18):9312. DOI: 10.3390/app12189312

[34] Gharibshahi R, Omidkhah M, Jafari A, Fakhroueian Z. Experimental investigation of nanofluid injection assisted microwave radiation for enhanced heavy oil recovery in a micromodel system. Korean Journal of Chemical Engineering. 2022;**39**(3):562-575. DOI: 10.1007/s11814-021-0961-7

[35] Druetta P, Raffa P, Picchioni F. Plenty of room at the bottom: Nanotechnology as solution to an old issue in enhanced oil recovery. Applied Sciences. 2018;**8**(12):2596. DOI: 10.3390/app8122596

[36] Hashemi R, Nassar NN, Almao PP. Nanoparticle technology for heavy oil in-situ upgrading and recovery enhancement: Opportunities and challenges. Applied Energy. 2014;**133**:374-387. DOI: 10.1016/j.apenergy.2014.07.069

[37] Cid A. Synthesis of NPs by microemulsion method. In: Microemulsion-a Chemical Nanoreactor. London, United Kingdom: IntechOpen; 2018. pp. 77-92. DOI: 10.5772/intechopen.80633.ch5

[38] Zielińska-Jurek A, Reszczyńska J, Grabowska E, Zaleska A. Nanoparticles preparation using microemulsion systems. In: Najjar R, editor. Microemulsions-an Introduction to Properties and Applications. Rijeka, Croatia: IntechOpen; 2012. pp. 229-250. DOI: 10.5772/36183.ch12

[39] Malik MA, Wani MY, Hashim MA. Microemulsion method: A novel route to synthesize organic and inorganic nanomaterials: 1st Nano update. Arabian Journal of Chemistry. 2012;**5**(4):397-417. DOI: 10.1016/j.arabjc.2010.09.027

[40] Abdrabo AE, Husein MM. Method for converting demetallization products into dispersed metal oxide

nanoparticles in heavy oil. Energy & Fuels. 2012;**26**(2):810-815. DOI: 10.1021/ef201819j

[41] Biyouki AA, Hosseinpour N, Bahramian A, Vatani A. In-situ upgrading of reservoir oils by in-situ preparation of NiO nanoparticles in thermal enhanced oil recovery processes. Colloids and Surfaces A: Physicochemical and Engineering Aspects. 2017;**520**:289-300. DOI: 10.1016/j.colsurfa.2017.01.089

[42] Mehrooz N, Gharibshahi R, Jafari A, Delavari H, Sadeghnejad S. Effect of In-situ synthesized cerium oxide nanoparticles on oil recovery factor in a glass micromodel. Applied Research in Chemical-Polymer Engineering. 2022;**6**(2):3-14

[43] Husein MM, Alkhaldi SJ. In situ preparation of alumina nanoparticles in heavy oil and their thermal cracking performance. Energy & Fuels. 2014;**28**(10):6563-6569. DOI: 10.1021/ef5012963

[44] Hashemi R, Nassar NN, Pereira Almao P. Enhanced heavy oil recovery by in situ prepared ultradispersed multimetallic nanoparticles: A study of hot fluid flooding for Athabasca bitumen recovery. Energy & Fuels. 2013;**27**(4):2194-2201. DOI: 10.1021/ef3020537

[45] Chen M, Li C, Li GR, Chen YL, Zhou CG. In situ preparation of well-dispersed CuO nanocatalysts in heavy oil for catalytic aquathermolysis. Petroleum Science. 2019;**16**(2):439-446. DOI: 10.1007/s12182-019-0300-3

[46] Tarboush BJA, Husein MM. Adsorption of asphaltenes from heavy oil onto in situ prepared NiO nanoparticles. Journal of Colloid and Interface Science.2012;**378**(1):64-69. DOI: 10.1016/j.jcis.2012.04.016

[47] Tarboush BJA, Husein MM. Dispersed Fe2O3 nanoparticles preparation in heavy oil and their uptake of asphaltenes. Fuel Processing Technology. 2015;**133**:120-127. DOI: 10.1016/j.fuproc.2014.12.049

[48] Hu Z, Nourafkan E, Gao H, Wen D. Microemulsions stabilized by in-situ synthesized nanoparticles for enhanced oil recovery. Fuel. 2017;**210**:272-281. DOI: 10.1016/j.fuel.2017.08.004

Chapter 5

Advancement in Hydraulic Fracturing for Improved Oil Recovery

Ahmed Merzoug, Habib Ouadi and Olusegun Tomomewo

Abstract

This chapter provides a comprehensive overview of advancements in hydraulic fracturing in unconventional plays. The narrative starts with an introduction to hydraulic fracturing and its transformative potential in the U.S., showcasing innovations in fracturing volumes, proppant masses, and well laterals. A detailed examination of fracturing fluids follows, emphasizing the dominance of slickwater treatments in unconventional plays. The chapter then delves into the crucial role of proppants, highlighting their surge in usage over a decade and the consequential shifts in material choice. The intricacies of perforation design are explored, particularly the revolutionary Xtreme Limited Entry approach and its subsequent impacts on production efficiency. In the realm of diagnostic technologies, the chapter presents a range, from traditional methods to emerging ones like Microseismic Depletion Delineation and time-lapse geochemical fingerprinting. The topic of refracturing is also addressed, spotlighting its merits in combating rapid production declines and the associated challenges. Finally, the chapter elucidates the phenomenon of fracture-driven interaction, offering insights into its historical context, influential factors, and proposed strategies to manage its repercussions. Through its breadth and depth, this chapter underscores the multifaceted nature of hydraulic fracturing advancements and their significance in the oil industry.

Keywords: hydraulic fracturing, unconventional plays, diagnostics, modeling, pilot studies

1. Introduction

In recent decades, the energy landscape has undergone transformative shifts, largely driven by advancements in drilling and extraction technologies [1]. Among these, hydraulic fracturing and horizontal drilling stand out as pivotal innovations that have revolutionized the field of unconventional reservoir exploitation in the United States [2]. These methodologies not only amplified production potentials but also ushered in a new era marked by constant technological advancements and optimizations. The upswing in hydraulic fracturing practices has catalyzed the inception of myriad new technologies, prompting in-depth, play-wide analyses that encompass a spectrum of reservoir characteristics and completion strategies [3]. Such analyses and innovations have borne fruit, with tangible increases in fracturing volumes, augmentation of

IntechOpen

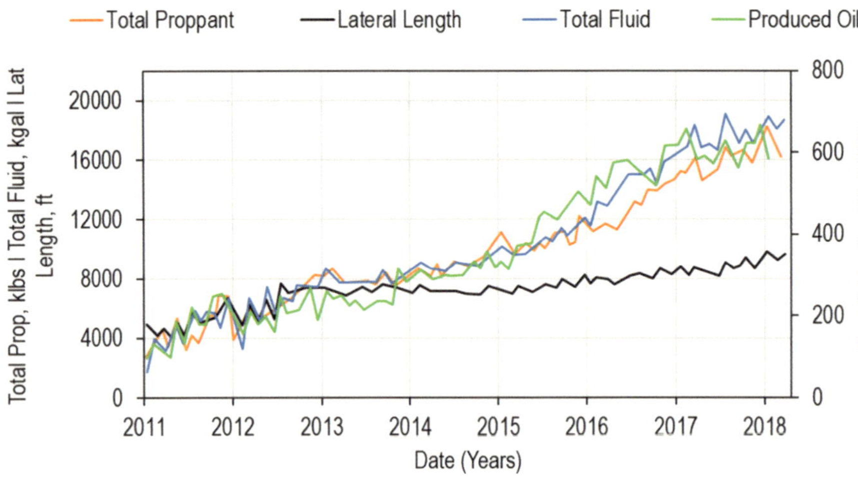

Figure 1.
Evolution of lateral length, total fluid volume, proppant mass, and oil production with time through the Midland Basin (modified from Xu et al. [5]).

proppant masses, and extensions of well laterals, leading to more efficient reservoir drainage and optimized production [4]. **Figure 1** offers a comprehensive visual representation of the progression of these completion parameters across a temporal spectrum. Within the ensuing sections of this chapter, readers will be guided through a detailed exploration of these advancements, understanding the mechanics, the methodologies, and their profound impacts on production. Through this synthesis, the chapter aims to provide a consolidated understanding of the current state of hydraulic fracturing, setting the stage for future innovations and research in the domain.

To navigate the intricate landscape of these advancements, this chapter unfolds as follows: We first delve deep into the design processes of hydraulic fracturing, tracing their evolution and significance in enhancing oil recovery. This exploration is supplemented by sub-sections focusing on the attributes of hydraulic fracturing fluids, the rising demand and adaptations of proppants, and the historical to modern developments in perforation design. Following this, we transition to a comprehensive examination of the diverse diagnostic techniques developed over the years, emphasizing their assumptions and limitations. The motivations, benefits, and challenges of refracturing existing wells form the subject of the subsequent section. We further explore the implications of fracture driven interaction in unconventional plays, discussing its history and strategies for mitigation. The chapter concludes by reflecting on the key findings and technological advancements, emphasizing their implications for future unconventional plays and the broader petroleum industry. Through this synthesis, we aim to provide a consolidated understanding of the current state of hydraulic fracturing, setting the stage for future innovations and research in the domain.

2. Completion design

2.1 Hydraulic fracturing fluids

The ideal fracturing fluid should meet several criteria. First, it must be compatible with both the formation materials and the fluids present. Additionally, it should have the ability to suspend and transport proppants deep into the fractures. Its inherent viscosity should be sufficient to create the required fracture width for proppant placement or for deep acid penetration. Efficiency is crucial, so the fluid should exhibit low leakoff. After its purpose is served, it should be readily removable from the formation. To ensure smooth operations, the fluid should generate minimal friction pressure and be straightforward to prepare on-site. Stability is another key characteristic; the fluid should maintain its viscosity throughout the treatment process. Lastly, while meeting all these criteria, it should also be cost-effective [6].

The use of slickwater (also called water frac or riverfrac) as the hydraulic fracturing treatment is the most popular approach in unconventional plays. This is mainly attributed to the simplicity of operations, ease of cleanup and lower cost [7]. On the other hand, the fluid has poor proppant transport properties due to its low viscosity. To overcome the transport challenges Zhao et al. [8] proposed a new type of fluid called High Viscosity Friction Reducer (HVFR). This new type of fluids has a low friction loss in the wellbore with good carrying capacity for proppant. This type of fluid system has gained a lot of popularity and have been applied in several basins to increase the proppant concentrations or increase the number of clusters per stage while ensuring good proppant transport capabilities [9–12].

2.2 Proppant

During the decade from 2010 to 2020, the surge in unconventional horizontal well completions led to a sharp increase in the amount of proppant used in North America, as noted by Weijers et al. [13]. This sharp rise in demand necessitated a substantial amount of proppant, prompting the use of readily available and cost-effective brown sand.

The early days of unconventional formations using horizontal wells initially leveraged learnings from the fracturing principles of conventional formations. In an effort to minimize costs, there were initial attempts to completely exclude proppant from the treatment [14]. However, these attempt were not sustainable in the long run. Coulter et al. [15] demonstrated that enhancing proppant quantities in unconventional horizontal well stimulations considerably boosted immediate production. This pattern of escalating proppant volumes per well has been aggressively adopted by the industry. As a result, the amount of proppant being used in many unconventional formations is now in the range of thousands of pounds per foot of lateral [13]. After the oil price plummeted in late 2014, the industry pivoted toward using more voluminous quantities of lower-grade proppants. Both Brown and White Sand proppants are now commonly used in formations with closure stresses reaching up to 10,000 psi [16, 17]. To emphasize on proppant selection importance Pearson et al. [17] showed a time dependent conductivity losses. They used both numerical simulation, laboratory experiements and field data to shwo the importance of good near wellbore conductivity. They noted that adding 5–10% lead and tail high condcutivty ceramic proppant can lead to a significant increase in production and higher return on investment. Singh et al. [18] leveraged field scale laboratory experiemnts and field trials to suggest a Constant Concentration (CoCo) propant pumping schedule. They applied the technique on more than 100 wells using low concentration proppant without using gel in the hydraulic fracturing fluids. They reported that their wells had similar normalized performance when compared to wells varying proppant concentration.

2.3 Perforation design

Limited entry was initially introduced by Murphy and Juch [19] and Lagrone and Rasmussen [20]. The primary purpose of this technique was to ensure consistent flow rates across various perforations at distinct breakdown pressures. Subsequently, this approach was adapted for unconventional reservoirs to address the subsequent challenges [21]:

- Variability in stress along the lateral length (with approximately 90% of laterals falling within the 750 psi range).
- Fluctuations in near-wellbore friction from one cluster to another (with a median value of 625 psi from step-down tests).
- Stress shadow effect between clusters and from preceding stages (influenced by formation elastic properties).
- Irregular fracture propagation within different clusters and the anisotropy of elastic properties.
- Alterations in perforation friction due to erosion and a range of perforation diameters (around 500 psi).

Note that the values in brackets are values measured for the Bakken [21]. The pressure drop through perforation can be calculated as follows [22]:

$$\Delta P_p = \frac{0.2369 \times Q^2 \times \rho}{N_p^2 \times D_p^2 \times C_d^2} \tag{1}$$

where ΔP_p, perforation pressure drop (pressure drop across perforation (psi)); Q, flow rate in bpm; ρ, fracturing fluid density (lb/gal); N_p, number of open perforations; D_p, perforation diameter (in); C_d, coefficient of discharge.

Weddle et al. [21] integrated the utilization of XLE (Xtreme Limited Entry) into their designs, leading to an increase in the cluster count from 11 clusters per stage to 15 clusters per stage, while maintaining a flow rate of 80 barrels per minute (bpm). The design involved a designated pressure drop of 2000 pounds per square inch (psi) at the perforations. The assessment of their achievements encompassed the analysis of data from fiber optic measurements, radioactive tracers, and production data.

For a lateral extent of 9500 feet, the authors managed to significantly reduce the total number of stages from 50 to 27, resulting in a pilot project that yielded an average incremental production increase of 10%. Notably, the authors highlighted the significance of considering plug shifting, which can induce leakage and subsequently compromise performance. The estimations related to proppant transport and settling velocity played a pivotal role in setting an upper limit on the number of perforation clusters feasible. Lorwongngam (Ohm) et al. [23] and Lorwongngam et al. [10] utilized eXtreem Limited Entry design to reduce the number of stages across the lateral to reduce the associated cost. They were able to increase the number of clusters per stage up to 15 then 20 with high unifromity index measured through Distributed Acoustic Sensing and downhole cameras.

To better understand the perforation design [24] conducted laboratory scale experiment. From their research, they concluded that perforation orientation has a significant effect on proppant placement. They reported that the response of proppant transprot is governed by a competion between viscous and gravity forces at lower rates, whereas at higher rates momentum effect become significant. Snider et al. [25] conducted field scale surface experiemnt. They reported that fluid placement and proppant placemnt are subject to different physics. Fluid placement is more uniform then proppant placement. With better unifromity for lower proppant size. The proppant distribution and errosion varied a lot in there testing. Dontsov [26] developed a mathemathical model to explain the physcis that governs proppant placement. He suggested that the proppant placement can explained by solving two subproblems: particle turning non unfirom proppant concentration in the wellbore. Using his mathemathical model he was able to match the previously mentioned experiments. He suggested that optimally the perforation orientation should be variable across the stage to ensure unifrom proppant placement; however since that is operationally challenging 90 degree orientation is the operationally optimum perforation orientation design [26].

3. Diagnostics technologies

Different fracture diagnostic techniques have been developed over the years to be able to characterize key fracture property. Each diagnostic approach has its own assumptions, limitations and key parameters. Barree et al. [27] described how to leverage early diagnostic techniques in evaluating the stimulation treatment. Microseismic can be used as a proxy for fracture geometry. As fractures propagate, they change the pressure distribution across the formation. This leads critically stressed natural fractures to slip generating a microseismic event. Fracture geometry (height, length, azimuth, and asymmetry) is then inferred from the combination of these events. Microseismic events generated the initial understanding of unconventional plays. The oil and gas industry used to think that generated system underground is a complex fracture network of fractures activated in different directions [28]. However, with the advancement of fracture diagnostic technology (Fiber optic), fractures were proven to be parallel plate fractures [29] as part of Hydraulic Fracturing Test Site (HFTS) project funded by the Department of Energy. Apparent discrepancies were noted at this project where microseismic events did not show fracture propagation, However the fracture propagation was felt by the strain on the offset well fiber optic.

Fiber optic was used in-well and offset wells for different purposes. The in-well fiber optic is used to evaluate the seal between two stages [23] or to quantify uniformity index between different clusters of the same stage [30, 31]. Jin et al. [32] utilized in-well fiber optic to quantify conductivity changes and fracture response during production. For offset well fiber optic usage, Dhuldhoya and Friehauf [33] used the technology for measuring far field uniformity and fracture length. Pudugramam et al. [34] used offset well fiber to estimate fracture height at the HFTS-2 project. Haustveit and Haffener [35] used fiber technology and bottom hole gauge to infer a relationship between pressure drawdown and fiber strain using polynomial function fitting.

Haustveit et al. [36] introduced Sealed Wellbore Pressure Monitoring as a low expense diagnostic approach to measure fracture driven interaction response at offset wells. It can be used for qualitatively assessing cluster efficiency, fluid distribution, estimate fracture height, and length. Identify depletion and estimate fracture closure time. It relies on monitoring deformation of a sealed wellbore. Once deformation occurs a pressure response is measured pressure gauge. The approach was validated using fiber optic in several basins [37, 38]. This technique was also used to calibrate fracture models using the volume to first response [39]. The volume to first response (VFR) is the volume injected before any pressure response is noted at offset wells. Smaller volumes are attributed to non-uniform fluid distribution as one fracture propagates faster than the others resulting in small VFR. This concept is illustrated in **Figure 2**. The left side of the figure depicts a high volume to first response, associated with a good distribution of fluids that delays the time for the first fracture hit to occur. Conversely, the right side shows a low volume to first response, indicating poor uniformity in fracture propagation. In this scenario, one fracture consumes more fluids and propagates, while other fractures remain stagnant.

Another diagnostic that leverage microseismic events is Microseismic Depletion Delineation (MDD). It was first introduced by Dohmen et al. [40]. The concept of the idea relies on poro-elastic stress changes due to depletion. As stress changes more the Mohr Columb failure criterion for different fractures shifts into critically stressed fractures. The operator injects a small volume of fluid to activate these critically stress fractures to map the stress state in the reservoir [41, 42]. The stress state is a proxy of the depletion areas. This concept is visually captured in **Figure 3**, which presents the Mohr-Coulomb failure criteria. The figure on the left illustrates the stress state after depletion, with optimally oriented natural fractures on the brink of slipping. However, upon water injection, the stress state undergoes a shift, as depicted in the figure on the right. This results in fracture slippage, generating microseismic events. These event are collocated at depleted zones. This approach have been applied several times to map the drainage area of parent wells in the Bakken and optimize well spacing of infill wells [43, 44].

Michael et al. [45] introduced a new diagnostics approach for mapping drained reservoir volume and dynamic production allocation by intervals through time in

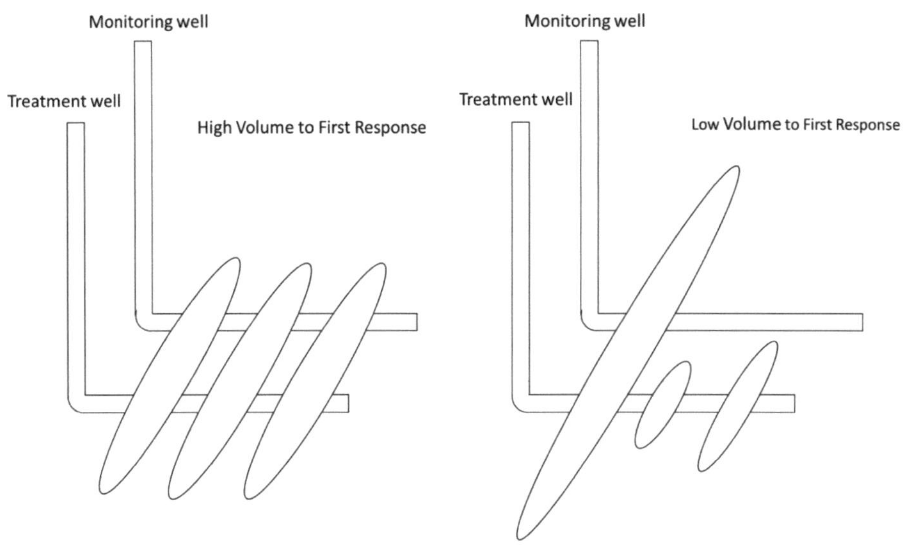

Figure 2.
Conceptual idea of first volume response modified from Haustveit et al. [38]. The figure in the left shows high volume to first response. This is associated with a good distribution of fluids delays the time for the first fracture hit to occur. On the right a low volume to first response illustrates a poor uniformity in fracture propagation as one fracture took more fluids and propagated whereas other fractures did not propagate.

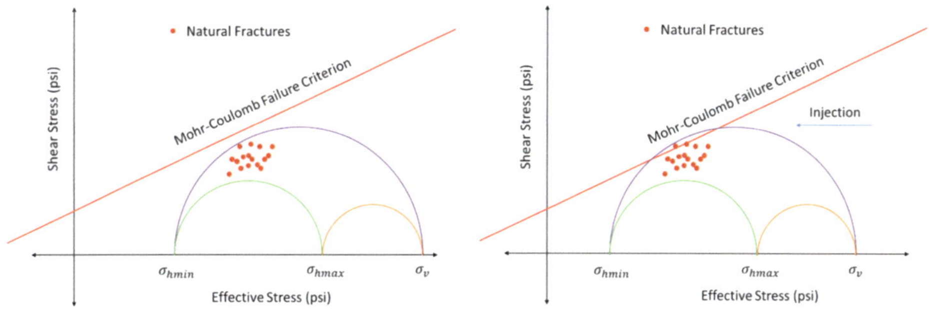

Figure 3.
Mohr-Coulomb failure criteria (left) stress state and natural fracture orientation after depletion, (right) stress state and fractures failure due to injection. The figure in the left shows the stress state after depletion. The natural fractures optimally oriented are about to slip. Once the water is injected the stress state shifts as illustrated in the figure in the right and the fractures slips generating microseismic events.

unconventional reservoirs. In their approach they leverage time-lapse geochemical (TLG) fingerprinting of formation fluids (gas, oil, and water). The approach relies on the unique biomarker signatures of each formation and their corresponding appearance in the produced fluids. This approach was used to optimize the landing depth and quantify the contribution of each member of the formation [46, 47]. Bachleda et al. [48] utilized the approach to build a proxy for EUR predictions in the Anadarko Basin. Maxwell et al. [49] leveraged the use of TLG and microseismic data to conceptually understand the effective drained fracture height.

Diagnostic Fracture Injection Test (DFIT) is a very popular diagnostics approach implemented in Unconventional plays. It consists of injecting small volumes into the wellbore to create a small fracture. The pressure/temperature response is then recorded to infer the formation least principal stress, formation permeability and formation pore pressure. Barree et al. [50] proposed an approach for the interpretation of DFIT tests. This approach have been widely used in the industry; however it have been criticized for the lack of physics based supporting evidence. McClure et al. [51, 52] identified issued with the holistic approach [50] and suggested a new interpretation scheme called the compliance method. McClure et al. [52] approach is specifically designed for low permeability formations and have been tested using field experiments and statistical approaches to outperform the holistic approach [53–56].

An interference test is a procedure in which one or more wells are sequentially brought into production, while the pressure is monitored in one or more nearby shut-in wells. By observing the pressure changes in the shut-in wells after each active well starts production, the degree of connectivity or communication between neighboring wells can be assessed. This type of tests is mainly used to determine optimum well spacing [57]. Well spacing decisions, both lateral and vertical, are fundamental for achieving economic efficiency in shale formations. While more densely spaced wells might result in reduced production and return on investment (ROI) per individual well, they can enhance the overall production and Net Present Value (NPV) for a specific land section [58].

Chu et al. [59] proposed an interpretation approach for interference tests. The approach have been applied in several basins [60, 61]. The approach quantifies a factor called Chow Pressure Group (CPG). The closer CPG is to 1 the higher the connectivity. The lower the CPG (less than 0.5) the lower the connectivity between wells [62]. Almasoodi et al. [63] reported that The CPG (Cross-Well Pressure Gradient) has certain limitations, including the inability to provide a quantitative estimate of how a well's production is affected by neighboring wells. Additionally, it is uncertain whether the same value of CPG, when measured between two wells, consistently indicates the same level of production interference, regardless of variables such as reservoir properties, fluid characteristics, or test conditions. They utilized analytical approach and numerical simulations to develop a physics-based interference interpretation approach (so-called Deveon Quantification Interference (DQI)). They have also applied the approach to different wells yielding consistent results.

4. Refracturing

Unconventional oil and gas wells frequently face rapid production declines. Historically, the industry's response has been to drill and fracture new wells. However, with recent oil price downturns, there's an escalating interest in refracturing, or restimulation, of existing wells to enhance their productivity. The objectives of refracturing are to increase the production of hydrocarbons from existing wells and improve their profitability [64]. This type of operations can result in an incremental production of 30–70% [65]. However, the success rate of refractured wells is relatively low, with only 15–20% of wells achieving the desired improvement in practice [65]. Therefore, one of the objectives of advanced studies in this area is to develop reliable and systematic approaches to increase the success rate of refracturing jobs.

Data analysis techniques and fuzzy clustering has been proposed for the selection of refracturing candidates, guiding the development of tight oil and gas reservoirs effectively [66]. Near-wellbore diversion is commonly used to ensure a uniform fluid and proppant distribution (particulate diversion, perforation sealing, and mechanical isolation [67]). The Barnett Shale has seen the application of multiple refracturing techniques, including bullhead treatment with and without diverter, various types of diversion, and mechanical isolation [68]. In Daqing Oilfield, refracturing has been employed to address the declining output of tight reservoirs. Different refracturing modes have been developed based on initial fracturing parameters and completions [69]. The Fuling shale gas field in China witnessed the first successful application of a Casing-in-Casing (CiC) refracturing treatment, providing a new option for refracturing horizontal wells in the region [70]. The use of biodegradable particulate diverters in hydraulic fracturing and refracturing has shown potential in increasing production and improving overall well economics, though their effectiveness can vary, especially in horizontal wells [71]. Another study on "blind" refracturing in horizontal wells in low-permeability reservoirs highlighted the potential of hybrid reinforcement, combining steel and basalt fiber-reinforced polymer (BFRP), which could increase the bearing capacity of samples by 33–74% [72].

Refracturing faces challenges such as limited options leading to increased costs and potential well loss. Current solutions include cementing perforations, expandable steel liners, biodegradable materials, and coiled tubing straddle packers, each with its own set of limitations. A promising approach is the reclosable sleeve technology, offering stage-by-stage access and maintaining wellbore integrity. These sleeves can be cycled open and closed during various well operations, providing flexibility in refracturing and production phases. However, their long-term effectiveness in refracturing remains to be fully validated [73]. The success of these refracturing techniques hinges on various factors, including the judicious selection of wells and determining the ideal timing for the procedure [65]. In reservoirs with natural fractures, understanding and predicting stress redistribution due to the poroelastic effect becomes intricate, making it crucial for optimizing refracturing performance [74]. Additionally, in areas like the Sulige Gasfield, wells with hydraulic fracturing have exhibited rapid decline rates, and the new fractures introduced during refracturing might not always align with the direction of the original fractures [75]. Additionally, selecting the optimal wells for refracturing is a primary challenge. Factors affecting the selection of refracturing candidates for multi-fractured horizontal wells are numerous and their relationships are complicated, making it difficult to choose the best wells [76]. The challenges of well refracturing include stage isolation, the complexity of the fracture system, candidate selection, understanding the rock system, refracture deviation, and uncertainty in predicting production results. These challenges highlight the need for careful planning, evaluation, and modeling to ensure the success of refracturing treatments.

5. Fracture driven interaction

To maximize the Estimated Ultimate Recovery (EUR) from unconventional plays, the design of hydraulic fracturing has observed tendency for larger fluid and proppant volumes and closer cluster spacing. However, these approaches led to a serious challenge known as fracture-driven interaction (FDI), also called frac bashing or frac-hit [77–79]. The first drilled well in unconventional is called primary or parent well, and the following drilled wells are called offset or child well. FDI refers to the inter-well communication between two wells.

The very first documented works in unconventional go back to Ajani and Kelkar [80] and Daneshy et al. [81]. Gupta et al. [82] reported a thorough literature review about different aspects of fracture driven interaction. They summarized different factors that influence the FDI response. The factors have been categorized into changeable and unchangeable. The changeable parameters are well placement, completion, and depletion; whereas several parameters are out of our control and are defined by the characteristics of the play. The unchangeable factors are divided into geological features and rock properties. The geological features consist of natural fractures, faults, bedding planes, fracture barriers, in situ stresses, rock fabric, and mineralogy. The rock properties consist of Young's modulus, Poisson's ratio, matrix permeability and porosity, fracture toughness, pore pressure, Biot's coefficient, and tensile strength.

Several authors worked on the influence of depletion on the occurrence of fracture driven interaction [43, 80, 82–86]. Depleting the primary well creates a low-stress area due to the poro-elastic response of the rock mass. When the fracture propagates from the offset well, a low-stress zone around the primary well causes an asymmetric growth of these fractures toward the lower stress area of the primary well [85, 87–89].

Jacobs [90] reported several completions strategies to avoid fracture driven interaction. He suggested the following practices with their limitations:

- Reduce the size of the offset well completion: This will reduce the capital invested and frac-hit occurrence; however, it can lead to an under-stimulated reservoir volume.

- Increase the size of the offset well completion: This will reduce the asymmetric growth of infill well fractures; however, it can lead to undercapitalization.

- On-the-Fly completions: also suggested by Daneshy [91], aims to adjust completion in the field according to measured data from the monitoring well. The difficulty is that this approach is technically challenging.
- Pinpoint/coil completion: It uses a single cluster to have more control over the fluid distribution; however, it can lead to smaller stimulated volume due to the limited number of clusters.
- Staggered wells, cube development, or rolling development can be implemented to either reduce the severity of the interaction or eliminate the depletion effect.

Fracture driven interaction was also investigated using numerical modeling approaches to understand the wells performance and how wells interact with each other's. Ratcliff et al. [92] modeled production losses due to fracture driven interaction in the Anadarko basin using a conductivity damage function. Fowler et al. [93] summarized the potential explanations for production uplifts [43, 94] and production losses [82]. McClure et al. [55, 56, 58] reported results from a cross basin collaborative study on modeling fracture driven interactions effect on production. They history matched several cases and suggested optimized scenarios for better wells development.

6. Challenges of hydraulic fracturing

The progression of hydraulic fracturing resulted in more questions then answers. Several areas of investigations are still open. The microseismic response explanation and interpretation are still ambiguous especially when compared to fiber optic [95]. The fracture driven interaction response in different plays [82] and their origin is another area of research to explain remediation approaches. Another challenge is the explanation of sub-parallel fractures that were noticed in [96]. Savitski [97] attempted to explain these responses; however, this area requires further investigation. One of the important challenges is the casing seal between stages. These isolation are provided through plugs and cement [98, 99]. Hydraulic fracturing propagation mechanisms in different setting is still a topic of discussion and several authors have reported a different way into solving the challenge [100]. This is still an area of active research. Furthermore, there is a room for proppant and completion selection criteria research to step in and utilize data analytics to improve prediction [101, 102].

7. Conclusion

The landscape of hydraulic fracturing has undergone a remarkable evolution over recent years, solidifying its role as a cornerstone in modern reservoir engineering practices, particularly in unconventional plays. From the nuances of fracturing fluid selection, which ensures compatibility and effective proppant transport, to the intricate science behind perforation design that optimizes production, we witness a confluence of innovations. The surge in proppant usage, and the resultant shifts in material choices, underscores the industry's adaptability to emerging challenges and market dynamics. Diagnostic technologies, with their breadth from early methodologies to novel techniques like Microseismic Depletion Delineation, stand testament to the relentless pursuit of more accurate reservoir characterizations. Refracturing emerges as a promising avenue to rejuvenate wells facing production declines, though it carries its own set of challenges which the industry is keenly addressing. Perhaps, one of the most significant takeaways from this chapter is the intricate dance of inter-well communication encapsulated by fracture-driven interaction. Its implications, both in terms of reservoir performance and economic outcomes, necessitate meticulous planning and agile execution strategies. As we move forward, it's evident that the industry's success will pivot on a deeper understanding of these elements, combined with a commitment to innovate and refine practices. The future of hydraulic fracturing, thus, promises to be as dynamic and transformative as its storied past.

Author details

Ahmed Merzoug[1*], Habib Ouadi[2] and Olusegun Tomomewo[2]

1 Texas A&M University, United States

2 University of North Dakota, United States

*Address all correspondence to: ahmed.merzoug@tamu.edu

References

[1] Pasqualetti M, Frantál B. The evolving energy landscapes of coal: Windows on the past and influences on the future. Moravian Geographical Reports. 2022;**30**(4):228-236. DOI: 10.2478/mgr-2022-0015

[2] Montgomery CT, Smith MB. Hydraulic fracturing: History of an enduring technology. Journal of Petroleum Technology. 2010;**62**(12):26-40. DOI: 10.2118/1210-0026-JPT

[3] Zhai M, Wang D, Li L, Zhang Z, Zhang L, Huang B, et al. Investigation on the mechanism of hydraulic fracture propagation and fracture geometry control in tight heterogeneous glutenites. Energy Exploration & Exploitation. 2021;**40**(1):246-278. DOI: 10.1177/01445987211036244

[4] Miller G, Lindsay G, Baihly J, Xu T. Parent well refracturing: Economic safety nets in an uneconomic market. In: Paper Presented at the SPE Low Perm Symposium Denver, Colorado, USA, May 2016. 2016. DOI: 10.2118/180200-ms

[5] Xu Y, Yu W, Sepehrnoori K. Development of an Efficient Method for Modeling Dynamic Fracture Behaviors in Reservoir Simulation. In: Paper Presented at the SPE/AAPG/SEG Unconventional Resources Technology Conference, Austin, Texas, USA, July 2017. 2018. DOI: 10.15530/urtec-2017-2670513

[6] Ely JW, Herndon RA. Fracturing fluids and additives. In: Miskimins J, editor. Hydraulic Fracturing: Fundamentals and Advancements. Society of Petroleum Engineers. Richardson, Texas, USA; 2020

[7] TTT P, MCC V, PJJ H. Slickwater fracturing: Food for thought. SPE Production & Operations. 2010;**25**(03):327-344. DOI: 10.2118/115766-PA

[8] Zhao H, Danican S, Torres H, Christanti Y, Nikolaev M, Makarychev-Mikhailov S, et al. Viscous slickwater as enabler for improved hydraulic fracturing design in unconventional reservoirs. In: Paper Presented at the SPE Annual Technical Conference and Exhibition, Dallas, Texas, USA, September, 2018. 2018. DOI: 10.2118/191520-MS

[9] Abdrazakov D, Juldugulov Y, Kruglov R, Pavlova S, Vereschagin S, Chuprakov D, et al. Proppant transport capabilities of high-viscosity friction reducers at relatively low rates: Advanced modeling and field validation. In: Paper SPE-203879-MS Presented at the Symposium: Hydraulic Fracturing in Russia, Virtual 22-24 September. 2020. DOI: 10.2118/203879-MS

[10] Lorwongngam AO, McKimmy M, Oughton E, Cipolla C. One shot wonder XLE design: A continuous improvement case study of developing XLE design in the Bakken. In: Paper Presented at the SPE Hydraulic Fracturing Technology Conference and Exhibition, The Woodlands, Texas, USA, January 2023. 2023. DOI: 10.2118/212358-MS

[11] O'Byrne K, de Groot M, Bonar D, Tkachyk J, Thiessen S. A Montney field study measuring the impact of completion designs on near and far field fracture geometry. In: Paper Presented at the SPE/AAPG/SEG Unconventional Resources Technology Conference, Denver, Colorado, USA, June 2023. 2023. DOI: 10.15530/urtec-2023-3857095

[12] Temizel C, Canbaz CH, Palabiyik Y, Hosgor FB, Atayev H, Ozyurtkan MH, et al. A review of hydraulic fracturing

and latest developments in unconventional reservoirs. In: Paper Presented at the Offshore Technology Conference, Houston, Texas, USA, May 2022. 2022. DOI: 10.4043/31942-MS

[13] Weijers L, Wright C, Mayerhofer M, Pearson M, Griffin L, Weddle P. Trends in the north American frac industry: Invention through the shale revolution. In: Paper Presented at the SPE Hydraulic Fracturing Technology Conference and Exhibition, The Woodlands, Texas, USA; 2019. DOI: 10.2118/194345-MS

[14] Mayerhofer MJ, Richardson MF, Walker RN, Meehan DN, Oehler MW, Browning RR. Proppants? We don't need no proppants. In: Paper Presented at the SPE Annual Technical Conference and Exhibition, San Antonio, Texas, October 1997. 1997. DOI: 10.2118/38611-MS

[15] Coulter GR, Benton EG, Thomson CL. Water fracs and sand quantity: A Barnett Shale example. In: Paper Presented at the SPE Annual Technical Conference and Exhibition, Houston, Texas, September 2004. 2004. DOI: 10.2118/90891-MS

[16] Melcher H, Mayerhofer M, Agarwal K, Lolon E, Oduba O, Murphy J, et al. Shale frac designs move to just-good-enough proppant economics. In: Paper Presented at the SPE Hydraulic Fracturing Technology Conference and Exhibition, The Woodlands, Texas, USA, February 2020. 2020. DOI: 10.2118/199751-MS

[17] Pearson CM, Fowler G, Stribling KM, McChesney J, McClure M, McGuigan T, et al. Near-wellbore deposition of high conductivity proppant to improve effective fracture conductivity and productivity of horizontal well stimulations. In: Paper Presented at the SPE Annual Technical Conference and Exhibition, Virtual, October 2020. 2020. DOI: 10.2118/201641-MS

[18] Singh A, Malhotra S, Wehunt D, Han S, Lannen C, Liu X, et al. Constant concentration proppant schedules for slickwater frac design in unconventional resources. In: Paper Presented at the SPE/AAPG/SEG Unconventional Resources Technology Conference, Houston, Texas, USA, July 2021. 2021. DOI: 10.15530/urtec-2021-5317

[19] Murphy WB, Juch AH. Pin-point sand fracturing—A method of simultaneous injection into selected sands. Journal of Petroleum Technology. 1960;**12**(11):21-24. DOI: 10.2118/1415-G

[20] Lagrone KW, Rasmussen JW. A new development in completion methods—The limited entry technique. Journal of Petroleum Technology. 1963;**15**(07):695-702. DOI: 10.2118/530-PA

[21] Weddle P, Griffin L, Pearson CM. Mining the Bakken II – Pushing the envelope with extreme limited entry perforating. In: Paper Presented at the SPE Hydraulic Fracturing Technology Conference and Exhibition, The Woodlands, Texas, USA, January 2018. 2018. DOI: 10.2118/189880-MS

[22] Cramer DD. The application of limited-entry techniques in massive hydraulic fracturing treatments. In: Paper Presented at the SPE Production Operations Symposium, Oklahoma City, Oklahoma, March 1987. Paper Number: SPE-16189-MS. 1987. DOI: 10.2118/16189-MS

[23] Lorwongngam (Ohm) A, Wright S, Hari S, Butler E, McKimmy M, Wolters J, et al. Using multidisciplinary data gathering to evaluate eXtreme Limited entry completion design and improve perforation cluster efficiency.

In: Paper Presented at the SPE/AAPG/SEG Unconventional Resources Technology Conference, Virtual, July 2020. Paper Number: URTEC-2020-2796-MS. 2020. DOI: 10.15530/urtec-2020-2796

[24] Ahmad F, Miskimins J, Liu X, Singh A, Wang J. Experimental investigation of proppant placement in multiple perforation clusters for horizontal fracturing applications. In: Paper Presented at the SPE/AAPG/SEG Unconventional Resources Technology Conference, Houston, Texas, USA, July 2021. 2021. DOI: 10.15530/urtec-2021-5298

[25] Snider P, Baumgartner S, Mayerhofer M, Woltz M. Execution and learnings from the first two surface tests replicating unconventional fracturing and proppant transport. In: Paper Presented at the SPE Hydraulic Fracturing Technology Conference and Exhibition, The Woodlands, Texas, USA, February 2022. 2022. DOI: 10.2118/209141-MS

[26] Dontsov E. A model for proppant dynamics in a perforated wellbore. International Journal of Multiphase Flow. 2023;**167**:104552. ISSN: 0301-9322. DOI: 10.1016/j.ijmultiphaseflow.2023.104552. Available from: http://arxiv.org/abs/2301.10855

[27] Barree RD, Fisher MK, Woodroof RA. A practical guide to hydraulic fracture diagnostic technologies. In: Paper Presented at the SPE Annual Technical Conference and Exhibition, San Antonio, Texas, September 2002. 2002. DOI: 10.2118/77442-MS

[28] Fisher MK, Heinze JR, Harris CD, Davidson BM, Wright CA, Dunn KP. Optimizing horizontal completion techniques in the barnett shale using microseismic fracture mapping. In: Paper Presented at the SPE Annual Technical Conference and Exhibition, Houston, Texas, September 2004. Paper Number: SPE-90051-MS. 2004. DOI: 10.2118/90051-MS

[29] Ugueto GA, Haffener J, Mondal S, Satviski AA, Huckabee PT, Haustveit K. Spatial and temporal effects on low frequency DAS and microseismic implications on hydraulic fracture geometry and well interactions. In: Paper Presented at the SPE Hydraulic Fracturing Technology Conference and Exhibition, The Woodlands, Texas, USA, February 2022. 2022. DOI: 10.2118/209180-MS

[30] Liang Y, Meier H, Srinivasan K, Tahir H, Ortega A, Amalokwu K, et al. Accelerating development optimization in the Bakken using an integrated fracture diagnostic pilot. In: Paper Presented at the SPE/AAPG/SEG Unconventional Resources Technology Conference, Houston, Texas, USA, June 2022. 2022. DOI: 10.15530/urtec-2022-3719696

[31] Lorwongngam AO, Cipolla C, Gradl C, Gil Cidoncha J, Davis B. Multidisciplinary data gathering to characterize hydraulic fracture performance and evaluate well spacing in the Bakken. In: Paper Presented at the SPE Hydraulic Fracturing Technology Conference and Exhibition, The Woodlands, Texas, USA, February 2019. 2019. DOI: 10.2118/194321-MS

[32] Jin G, Ugueto G, Wojtaszek M, Guzik A, Jurick D, Kishida K. Novel near-wellbore fracture diagnosis for unconventional Wells using high-resolution distributed strain sensing during production. SPE Journal. 2021;**26**(05):3255-3264. DOI: 10.2118/205394-PA

[33] Dhuldhoya K, Friehauf K. Applications of distributed strain sensing

via Rayleigh frequency shift: Illuminating near-well and far-field fracture characteristics. In: Paper Presented at the SPE/AAPG/SEG Unconventional Resources Technology Conference, Houston, Texas, USA, June 2022. 2022. DOI: 10.15530/urtec-2022-3721749

[34] Pudugramam VS, Zhao Y, Bessa F, Li J, Zakhour N, Brown T, et al. Analysis and integration of the hydraulic fracturing test Site-2 (HFTS-2) comprehensive dataset. In: Paper Presented at the SPE/AAPG/SEG Unconventional Resources Technology Conference, Houston, Texas, USA, July 2021. 2021. DOI: 10.15530/urtec-2021-5241

[35] Haustveit K, Haffener J. Can you feel the pressure? Strain-based pressure estimates. In: Paper Presented at the SPE Hydraulic Fracturing Technology Conference and Exhibition, The Woodlands, Texas, USA, January 2023. 2023. DOI: 10.2118/212364-MS

[36] Haustveit K, Elliott B, Haffener J, Ketter C, O'Brien J, Almasoodi M, et al. Monitoring the pulse of a well through sealed wellbore pressure monitoring, a breakthrough diagnostic with a multi-basin case study. In: Paper Presented at the SPE Hydraulic Fracturing Technology Conference and Exhibition, The Woodlands, Texas, USA, February 2020. 2020. DOI: 10.2118/199731-MS

[37] Cipolla C, McKimmy M, Hari-Roy S, Wolters J, Haffener J, Haustveit K, et al. Evaluating stimulation effectiveness with permanent optical fiber and sealed wellbore pressure monitoring: Bakken case study. In: Paper Presented at the SPE Hydraulic Fracturing Technology Conference and Exhibition, February 1-3, 2022. 2022. DOI: 10.2118/209129-MS

[38] Haustveit K, Elliott B, Haffener J, Ketter C, O'Brien J, Almasoodi M, et al. Monitoring the pulse of a well through sealed wellbore pressure monitoring, a breakthrough diagnostic with a multi-basin case study. In: Paper Presented at the SPE Hydraulic Fracturing Technology Conference and Exhibition, The Woodlands, Texas, USA, February 2020. 2020. DOI: 10.2118/199731-ms

[39] Olson K, Merritt J, Barraez R, Fowler G, Haffener J, Haustveit K. Sealed wellbore pressure monitoring (SWPM) and calibrated fracture modeling: The next step in unconventional completions optimization. In: Paper Presented at the SPE Hydraulic Fracturing Technology Conference and Exhibition, The Woodlands, Texas, USA, January 2023. 2023. DOI: 10.2118/212367-MS

[40] Dohmen T, Zhang J, Li C, Blangy JP, Simon KM, Valleau DN, et al. A new surveillance method for delineation of depletion using microseismic and its application to development of unconventional reservoirs. Proceedings - SPE Annual Technical Conference and Exhibition. 2013;**3**:2373-2386. DOI: 10.2118/166274-ms

[41] Dohmen T, Zhang J, Barker L, Blangy JP. Microseismic magnitudes and b -values for delineating hydraulic fracturing and depletion. SPE Journal. 2017;**22**(05):1624-1634. DOI: 10.2118/186096-PA

[42] Dohmen T, Blangy J-P, Zhang J. Microseismic depletion delineation. Interpretation. 2014;**2**(3):SG1-SG13. DOI: 10.1190/INT-2013-0164.1

[43] Cipolla C, Litvak M, Prasad RS, McClure M. Case history of drainage mapping and effective fracture length in the Bakken. In: Paper Presented at the SPE Hydraulic Fracturing Technology Conference and Exhibition, The Woodlands, Texas, USA, February 2020. 2020. pp. 1-43. DOI: 10.2118/199716-ms

[44] Lorwongngam AO, Cipolla C, Gradl C, Cidoncha JG, Davis B. Multidisciplinary data gathering to characterize hydraulic fracture performance and evaluate well spacing in the Bakken. In: Paper Presented at the SPE Hydraulic Fracturing Technology Conference and Exhibition, The Woodlands, Texas, USA, February 2019. 2019. DOI: 10.2118/194321-ms

[45] Michael G, Liu F, Brown D, Johansen K. Time lapse geochemistry application in unconventional reservoir development. In: Paper Presented at the SPE/AAPG/SEG Unconventional Resources Technology Conference, Austin, Texas, USA, July 2017. 2017. DOI: 10.15530/urtec-2017-2670186

[46] Bachleda J, Call K, Hardman D, Montoya E, Jin M, Wu J, et al. Optimizing development of stacked reservoirs and identifying depleted reservoir intervals using geochemical data. In: Paper Presented at the SPE/AAPG/SEG Unconventional Resources Technology Conference, Denver, Colorado, USA, June 2023. 2023. DOI: 10.15530/urtec-2023-3865769

[47] Jin M, Esmaili S, Harris A, Pavlovic M, Serna A, Lv T, et al. Reservoir characterization and monitoring using geochemical fingerprinting technology with case studies in the Delaware Basin. In: Paper Presented at the SPE/AAPG/SEG Unconventional Resources Technology Conference, Houston, Texas, USA, June 2022. 2022. DOI: 10.15530/urtec-2022-3723632

[48] Bachleda J, Norton G, Kennedy W, Lv T, Wu J, Liu F. Reliable EUR prediction using geochemistry-derived drainage profiles of 200+ wells in the Anadarko Basin. In: Paper Presented at the SPE/AAPG/SEG Unconventional Resources Technology Conference, Houston, Texas, USA, June 2022. 2022. DOI: 10.15530/urtec-2022-3722977

[49] Maxwell S, Brito R, Ritter G, Sinclair J, Leavitt A, Liu F, et al. Evaluation of effective drainage height through integration of microseismic and geochemical depth profiling of produced hydrocarbons. In: Paper Presented at the SPE Hydraulic Fracturing Technology Conference and Exhibition, The Woodlands, Texas, USA, January 2023. 2023. DOI: 10.2118/212314-MS

[50] Barree RD, Barree VL, Craig DP. Holistic fracture diagnostics. In: Paper Presented at the Rocky Mountain Oil & Gas Technology Symposium, Denver, Colorado, USA, April 2007. 2007. DOI: 10.2118/107877-MS

[51] McClure MW, Biyton CAJ, Jung H, Sharma MM. The effect of changing fracture compliance on pressure transient behavior during diagnostic fracture injection tests. In: Paper Presented at the SPE Annual Technical Conference and Exhibition, Amsterdam, The Netherlands, October 2014. Vol. 7. 2014. pp. 4973-4995. DOI: 10.2118/170956-ms

[52] McClure M, Bammidi V, Cipolla C, Cramer D, Martin L, Savitski A, et al. A collaborative study on DFIT interpretation: Integrating modeling, field data, and analytical techniques. In: Paper Presented at the SPE/AAPG/SEG Unconventional Resources Technology Conference, Denver, Colorado, USA, July 2019. 2019. pp. 1-39. DOI: 10.15530/urtec-2019-123

[53] Guglielmi YMM, Burghardt J, Joseph PM, Doe T, Fu P, Knox H. Estimating stress from fracture injection tests: Comparing pressure transient interpretations with in-situ strain measurements. In: Proceedings, 47th Workshop on Geothermal Reservoir Engineering, Stanford University,

Stanford, California, February 7-9, 2022. 2022. pp. 1-19. SGP-TR-22

[54] Guglielmi Y, McClure M, Burghardt J, Morris JP, Doe T, Fu P, et al. Using in-situ strain measurements to evaluate the accuracy of stress estimation procedures from fracture injection/shut-in tests. International Journal of Rock Mechanics and Mining Sciences. 2023;**170**:105521. DOI: 10.1016/j.ijrmms.2023.105521

[55] McClure M, Albrecht M, Bernet C, Etcheverry K, Fuhr A, Gherabati A, et al. Results from a collaborative parent/child industry study: Permian Basin. In: Paper Presented at the SPE 2022 Symposium Compilation, Virtual, December 2022. 2022. DOI: 10.2118/211899-MS

[56] Mcclure M, Fowler G, Picone M. Best practices in DFIT interpretation: Comparative analysis of 62 DFITs from nine different shale plays. In: Paper Presented at the SPE International Hydraulic Fracturing Technology Conference & Exhibition, Muscat, Oman, January 2022. 2022. DOI: 10.2118/205297-MS

[57] Almasoodi M, Vaidya R, Reza Z. Intra-well interference in tight oil reservoirs: What do we need to consider? Case study from the meramec. In: Unconventional Resources Technology Conference, Denver, Colorado, 22-24 July 2019. pp. 1-22. DOI: 10.15530/urtec-2019-83

[58] McClure M, Albrecht M, Bernet C, Cipolla C, Etcheverry K, Fowler G, et al. Results from a collaborative industry study on parent/child interactions: Bakken, Permian Basin, and Montney. In: Paper Presented at the SPE Hydraulic Fracturing Technology Conference and Exhibition, The Woodlands, Texas, USA, January 2023. 2023. DOI: 10.2118/212321-MS

[59] Chu W-C, Scott KD, Flumerfelt R, Chen C, Zuber MD. A new technique for quantifying pressure interference in fractured horizontal shale Wells. SPE Reservoir Evaluation & Engineering. 17 Feb 2020;**23**(01):143-157. DOI: 10.2118/191407-PA

[60] Miranda C, Cipolla C, Murray M. Analysis of well interference tests in the Bakken shale, a fully communicated system. In: Paper Presented at the SPE/AAPG/SEG Unconventional Resources Technology Conference, Houston, Texas, USA, June 2022. 2022. DOI: 10.15530/urtec-2022-3717749

[61] Petter Z, Warren C, Paz R, Bacon J, Carter A, Stang B, et al. Operational application of chow pressure group methodology in the Midland Basin and resulting development impact. In: Paper Presented at the SPE/AAPG/SEG Unconventional Resources Technology Conference, Denver, Colorado, USA, June 2023. 2023. DOI: 10.15530/urtec-2023-3855890

[62] Ballinger B, Green B, Vajjha P, Haffener J, Edwards M, Almasoodi M, et al. Understanding the hydraulic and conductive half lengths in the bakken/three forks play – coupling sealed wellbore pressure monitoring SWPM & chow pressure group CPG. In: Paper Presented at the SPE Hydraulic Fracturing Technology Conference and Exhibition, The Woodlands, Texas, USA, February 2022. 2022. DOI: 10.2118/209144-MS

[63] Almasoodi M, Andrews T, Johnston C, Singh A, McClure M. A new method for interpreting well-to-well interference tests and quantifying the magnitude of production impact: Theory and applications in a multi-basin case study. Geomechanics and Geophysics for Geo-Energy and Geo-Resources. 2023;**9**(1):95. DOI: 10.1007/s40948-023-00632-1

[64] Cafaro D, Drouven M, Grossmann I. Optimization models for planning shale gas well refracture treatments. AICHE Journal. 2016;**62**(12):4297-4307. DOI: 10.1002/aic.15330

[65] Tavassoli S, Yu W, Javadpour F, Sepehrnoori K. Well screen and optimal time of refracturing: A Barnett shale well. Journal of Petroleum Engineering. 18 Apr 2013;**2013**:1-10. DOI: 10.1155/2013/817293

[66] Tao L, Guo J, Zhao Z, et al. Refracturing candidate selection for MFHWs in tight oil and gas reservoirs using hybrid method with data analysis techniques and fuzzy clustering. Journal of Central South University. 2020;**27**:277-287. DOI: 10.1007/s11771-020-4295-0

[67] Zhang J, White M, McEwen J, Schroeder S, Cramer DD. Investigating near-wellbore diversion methods for refracturing horizontal Wells. In: Paper Presented at the SPE Hydraulic Fracturing Technology Conference and Exhibition, The Woodlands, Texas, USA, February 2020. 2020. DOI: 10.2118/199703-MS

[68] Senters C, Brady J, Werline R. A diagnostic evaluation of refracturing techniques. In: Paper Presented at the SPE Annual Technical Conference and Exhibition, Calgary, Alberta, Canada, September 2019. 2019. DOI: 10.2118/196195-MS

[69] Wang X, Wang W, Li B, Li D, Gu M, Tang P, et al. The application of refracturing technique in horizontal Wells of Daqing oilfield. In: Paper Presented at the Abu Dhabi International Petroleum Exhibition & Conference, Abu Dhabi, UAE, November 2018. 2018. DOI: 10.2118/192787-MS

[70] Liu Y, Li Y, Zhang C, Ming Y, Xiao J, Han R, et al. First successful application of casing in casing CiC refracturing treatment in shale gas well in China: Case study. In: Paper Presented at the Abu Dhabi International Petroleum Exhibition & Conference; Abu Dhabi, UAE. 2021. DOI: 10.2118/208060-MS

[71] Akrad O, Miskimins J. The use of biodegradable particulate diverting agents in hydraulic fracturing and refracturing: An experimental study. In: Paper Presented at the International Petroleum Technology Conference; Bangkok, Thailand. 2023. DOI: 10.2523/IPTC-23059-MS

[72] Sadykov AM, Erastov SA, Antonov MS, Kashapov DV, Salakhov TR, Kardopoltsev AS, et al. Scientific approach to planning and implementation of blind refracturing in horizontal wells with MSF completion in low-permeability reservoirs. Paper Presented at the SPE Russian Petroleum Technology Conference. 2021. DOI: 10.2118/206406-MS

[73] Phillips J. Refrac Facts: Technology Key to Enabling Refracs. The American Oil & Gas Reporter; 2022. Available from: https://www.aogr.com/magazine/sneak-peek-preview/technology-key-to-enabling-refracs [Accessed: August 23, 2023]

[74] Sangnimnuan A, Li J, Wu K, Holditch SA. Application of efficiently coupled fluid flow and geomechanics model for refracturing in highly fractured reservoirs. Paper Presented at the SPE Hydraulic Fracturing Technology Conference and Exhibition; The Woodlands, Texas, USA. 2018. DOI: 10.2118/189870-MS

[75] Wang Y, Yao Y, Wang L, Hu Y, Wu H, Wang H. Case study: Analysis of refracturing crack orientation-angle and extension-length in tight gas reservoir, Sulige Gasfield of China.

In: Paper Presented at the SPE EuropEC - Europe Energy Conference Featured at the 83rd EAGE Annual Conference & Exhibition; Madrid, Spain. 2022. DOI: 10.2118/209633-MS

[76] Tao L, Zhang X, Yang S, Xian S, Zhao Y, Zhao Z. A new re-fracturing candidate selection method for multi-fractured horizontal well in tight oil reservoirs. Paper Presented at the IADC/SPE Asia Pacific Drilling Technology Conference, Virtual. 2021. DOI: 10.2118/201059-MS

[77] Gupta I, Rai C, Devegowda D, Sondergeld C. A data-driven approach to detect and quantify the impact of frac-hits on parent and child wells in unconventional formations. In: Paper Unconventional Resources Technology Conference, Virtual, 20-22 July 2020. 2020. pp. 71-85. DOI: 10.15530/urtec-2020-2190

[78] Xu T, Lindsay G, Zheng W, Yan Q, Patron KE, Alimahomed F, et al. Advanced modeling of production induced pressure depletion and well spacing impact on infill wells in spraberry, Permian basin. In: Paper Presented at the SPE Annual Technical Conference and Exhibition, Dallas, Texas, USA, September 2018. 2018. DOI: 10.2118/191696-ms

[79] Yan M, Rahnema H, Shabani B. Application of machine learning to study frac-hit. Society of Petroleum Engineers - SPE Oklahoma City Oil and Gas Symposium 2019, OKOG 20192019. DOI: 10.2118/195187-ms

[80] Ajani A, Kelkar M. Interference study in shale plays. In: Paper Presented at the SPE Hydraulic Fracturing Technology Conference, The Woodlands, Texas, USA, February 2012. 2012. pp. 36-50. DOI: 10.2118/151045-ms

[81] Daneshy A, Au-Yeung J, Thompson T, Tymko D. Fracture shadowing: A direct method for determining of the reach and propagation pattern of hydraulic fractures in horizontal wells. In: Paper Presented at the SPE Hydraulic Fracturing Technology Conference, The Woodlands, Texas, USA, February 2012. 2012. pp. 215-223. DOI: 10.2118/151980-ms

[82] Gupta I, Rai C, Devegowda D, Sondergeld CH. Fracture hits in unconventional reservoirs: A critical review. SPE Journal. 10 Feb 2021;**26**(1): 412-434. DOI: 10.2118/203839-PA

[83] Gala DP, Manchanda R, Sharma MM. Modeling of fluid injection in depleted parent wells to minimize damage due to frac-hits. In: Paper Presented at the SPE/AAPG/SEG Unconventional Resources Technology Conference, Houston, Texas, USA, July 2018. 2018. pp. 1-14. DOI: 10.15530/urtec-2018-2881265

[84] Manchanda R, Bhardwaj P, Hwang J, Sharma MM. Parent-child fracture interference: Explanation and mitigation of child well underperformance. In: Paper Presented at the SPE Hydraulic Fracturing Technology Conference and Exhibition, The Woodlands, Texas, USA, January 2018. 2018. DOI: 10.2118/189849-ms

[85] Rezaei A, Dindoruk B, Soliman MY. On parameters affecting the propagation of hydraulic fractures from infill wells. Journal of Petroleum Science and Engineering. 2019;**182**(July):106255. DOI: 10.1016/j.petrol.2019.106255

[86] Zheng S, Manchanda R, Gala D, Sharma M, Hwang J. Pre-loading depleted parent wells to avoid frac-hits: Some important design considerations. In: Paper Presented at the SPE Annual Technical Conference and Exhibition,

Calgary, Alberta, Canada, September 2019. 2019. DOI: 10.2118/195912-ms

[87] Gupta JK, Zielonka MG, Albert RA, El-Rabaa W, Burnham HA, Choi NH. Integrated methodology for optimizing development of unconventional gas resources. In: Paper Presented at the SPE Hydraulic Fracturing Technology Conference, The Woodlands, Texas, USA, February 2012. 2012. pp. 507-522. DOI: 10.2118/152224-ms

[88] Kumar D, Ghassemi A, Riley S, Elliott B. Geomechanical analysis of frac-hits using a 3D poroelastic hydraulic fracture model. In: Paper Presented at the SPE Annual Technical Conference and Exhibition, Dallas, Texas, USA, September 2018. 2018. DOI: 10.2118/191491-ms

[89] Morsy S, Menconi M, Liang B. Unconventional Optimized Development Strategy Workflow. In: Paper Presented at the SPE Western Regional Meeting, Virtual, April 2021. Vol. 2021. May 2020. pp. 1-9. DOI: 10.2118/200775-MS

[90] Jacobs T. To solve frac hits, unconventional engineering must revolve around them. Journal of Petroleum Technology. 1 Apr 2019;**71**(04):27-31. DOI: 10.2118/0419-0027-jpt

[91] Daneshy A. Frac-Driven Interactions (FDI) Guides for Real-Time Analysis & Execution of Fracturing Treatment. 2020

[92] Ratcliff D, McClure M, Fowler G, Elliot B, Qualls A. Modelling of parent child well interactions. In: Paper Presented at the SPE Hydraulic Fracturing Technology Conference and Exhibition. The Woodlands, Texas, USA, February 2022. 2022. DOI: 10.2118/209152-ms

[93] Fowler G, Ratcliff D, McClure M. Modeling frac hits: Mechanisms for damage versus uplift. In: Paper Presented at the International Petroleum Technology Conference, Riyadh, Saudi Arabia, February 2022. 2022. DOI: 10.2523/IPTC-22194-MS

[94] Fowler G, McClure M, Cipolla C. Making sense out of a complicated parent/child well dataset: A Bakken case study. In: Paper Presented at the SPE Annual Technical Conference and Exhibition, Virtual, October 2020. 2020. DOI: 10.2118/201566-ms

[95] Ugueto GA, Wojtaszek M, Huckabee PT, Savitski AA, Guzik A, Jin G, et al. An integrated view of hydraulic induced fracture geometry in hydraulic fracture test site 2. In: Paper Presented at the SPE/AAPG/SEG Unconventional Resources Technology Conference, Houston, Texas, USA, July 2021. 2021. DOI: 10.15530/urtec-2021-5396

[96] Raterman KT, Farrell HE, Mora OS, Janssen AL, Busetti S, McEwan J, et al. Sampling a stimulated rock volume: An eagle ford example. In: Paper Presented at the SPE/AAPG/SEG Unconventional Resources Technology Conference, Austin, Texas, USA, July 2017. 2017. DOI: 10.15530/urtec-2017-2670034

[97] Savitski AA. On a poroelastic nature of multi-stranded hydraulic fractures in heterogeneous unconventional rocks. In: Paper Presented at the 54th U.S. Rock Mechanics/Geomechanics Symposium. Physical event canceled, June 2020

[98] White M, Friehauf K, Cramer D, Constantine J, Zhang J, Schmidt S, et al. One stage forward or two stages back: What are we treating? Identification of internal casing erosion during hydraulic fracturing—A montney case study using ultrasonic and fiber-optic diagnostics. SPE Drilling & Completion. 2021;**36**(02):383-397. DOI: 10.2118/201734-PA

[99] Zheng C, Liu Z-K, Wu X-L, Abulimiti A, Qin J, Zheng X-F, et al. Effect of structural parameters on the setting performance of plug slips during hydraulic fracturing. Petroleum Science. 2022;**19**(2):731-742. DOI: 10.1016/j.petsci.2021.09.011

[100] Chen B, Barboza BR, Sun Y, Bai J, Thomas HR, Dutko M, et al. A review of hydraulic fracturing simulation. In: Archives of Computational Methods in Engineering. Vol. 29(4). Netherlands: Springer; 2022. DOI: 10.1007/s11831-021-09653-z

[101] Zoveidavianpoor M, Gharibi A. Applications of type-2 fuzzy logic system: Handling the uncertainty associated with candidate-well selection for hydraulic fracturing. Neural Computing and Applications. 2016;**27**(7):1831-1851. DOI: 10.1007/s00521-015-1977-x

[102] Zoveidavianpoor M, Gharibi A, Bin Jaafar MZ. Experimental characterization of a new high-strength ultra-lightweight composite proppant derived from renewable resources. Journal of Petroleum Science and Engineering. 2018;**170**:1038-1047. DOI: 10.1016/j.petrol.2018.06.030

Chapter 6

In-Situ Foam Generation: A Superior Method for Enhanced Oil Recovery in Unconventional Fractured Reservoirs

Magda Ibrahim Youssif

Abstract

Unconventional reservoirs, like shale gas, shale oil, tight gas sands, and coalbed methane deposits, pose unique challenges due to their low permeability, low porosity, and complex geological structures. These factors hinder the natural flow of hydrocarbons, necessitating advanced extraction techniques. Hydraulic fracturing is commonly used to increase permeability and enhance hydrocarbon recovery. However, this creates a challenge during gas injection due to significant permeability differences between fractures and matrix. Foam flooding is an innovative enhanced oil recovery method in heterogeneous systems. It reduces fracture transmissivity and improves matrix-fracture interactions, thus enhancing oil sweep efficiency. Yet, foam stability depends on the method of generation. Traditional foam pre-generation at the surface is ineffective in fractured systems as foam loses its properties during transport under high pressure and temperature. This study's primary objective is to develop in-situ foam generation under reservoir conditions within fractured systems to enhance oil displacement. Achieving this involves optimizing factors like surfactant formulation, concentration, injection rate, and gas fraction. Additionally, the reservoir's petrophysical properties like wettability, permeability, and mineral composition, are considered. As a result of these efforts, the foam generated in situ will possess the capability to adapt to prevailing conditions and boost hydrocarbon production from such reservoirs.

Keywords: unconventional reservoir, hydraulic fracturing, enhanced oil recovery, gas mobility control, In-situ foam flooding, foam stability, apparent viscosity

1. Introduction

Unconventional formations all over the world have emerged as crucial contributors to hydrocarbon recovery [1]. However, extracting oil efficiently from these reservoirs remains a significant challenge using the current techniques. The unique characteristics of unconventional formations pose obstacles to efficient oil production. Unlike conventional reservoirs with high permeability and well-connected pore networks, unconventional formations have low permeability and are typically composed of tight rock, such as shale or sandstone. The low permeability restricts the flow of oil, gas, and other fluids, making it more difficult to extract hydrocarbons [2–4]. In addition to low permeability, unconventional reservoirs often exhibit complex geology and heterogeneous properties [5]. The rock formations can have variations in porosity, mineral composition, and organic content, which further complicate the extraction process [6].

To overcome some of these challenges, innovative research has developed and implemented various technologies and techniques. Hydraulic fracturing, commonly known as fracking, has revolutionized the extraction of oil and gas from unconventional formations [7]. This process involves injecting high-pressure fluid into the reservoir to create fractures in the rock, thereby enhancing permeability and allowing hydrocarbons to flow more freely [8]. However, when hydraulic fractures are induced in specific regions of a reservoir, it can introduce a higher degree of complexity to the pore structure. This complexity is characterized by variations in permeability, creating a contrast between the fracture and the surrounding matrix [9]. In other words, the permeability of the fracture, which is the ease with which fluids can flow through it, can differ significantly from that of the adjacent matrix rock. This contrast in permeability between the fracture and matrix poses a challenge to efficient fluid flow and can impact the effectiveness of secondary recovery and/or enhanced oil recovery (EOR) techniques in such complex reservoir systems [9, 10]. For instance, the major challenge arising during secondary gas injection in fractured reservoirs is its poor volumetric sweep efficiency, because gas tends to flow through high-permeability streaks rather than smaller ones. As a result, gas breakthrough occurs, bypassing the resident fluids and ultimately reaching the production wells [11–15]. Moreover, owing to the high gas mobility, gas tends to override the resident fluids in place resulting in poor volumetric sweep efficiency as shown in **Figure 1** [11, 17]. Consequently, a large fraction of oil does not come in contact with gas and the overall recovery remains low.

In order to mitigate such challenges in fractured reservoirs, the foam-based EOR technique addresses the drawbacks associated with gas injection by increasing the viscous pressure gradient through foam generation and therefore, reducing gas mobility in fractures and enhancing volumetric sweep efficiency [18–21].

In the upcoming sections, we will discuss the fundamental principles of foam transport in porous media. In addition, the methods used to measure foam texture and its stability are discussed. Besides, the standard and innovative approaches for foam generation in fractured reservoirs are also highlighted. Summary and recommendations of foam applications in such reservoirs are listed along with the conclusions emphasizing the broader impacts and significance of foam EOR.

2. Foam fundamentals

Simulating foam transport in porous media is a complex task that requires advanced numerical models and experimental data to accurately predict foam behavior in different reservoir conditions. It is imperative to draw insights from laboratory investigations, core flooding experiments, and reservoir simulations. These combined efforts yield a dynamic and efficient process for in-situ foam generation, specifically tailored to the complexities of porous media found in reservoirs. When considering the concept of on-site foam generation within reservoirs, several foundational principles must be incorporated to unravel the distinct attributes of foam and validate its efficacy in enhancing hydrocarbon displacement within the reservoir. This approach involves a thorough understanding of how a foaming agent, such as a surfactant

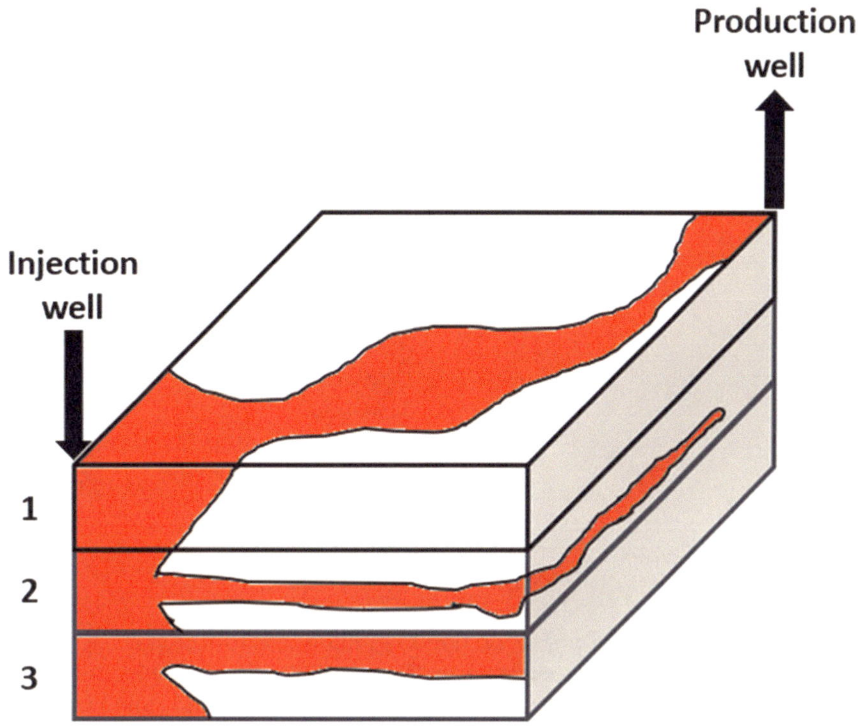

Figure 1.
Problems encountered by gas flooding in a fractured system (1) gas breakthrough due to high-permeability contrast, (2) gas channeling owing to high gas mobility, and (3) gas overriding due to low density [16].

solution, operates by minimizing surface tension at the gas-water interface, thereby encapsulating gas bubbles. Subsequently, a comprehensive exploration of foam strength becomes crucial, gauged by two key parameters: foam viscosity and the mobility reduction factor (MRF). These metrics essentially gauge the foam's capacity to regulate gas mobility within a fractured medium. For instance, a stable foam possessing high MRF and viscosity exerts significant control over gas mobility within fractures, redirecting it toward less permeable matrix zones to access oil-filled areas that would otherwise remain untouched. This mechanism substantially amplifies fracture-matrix interactions due to the adept gas mobility management achieved through the application of foam. Furthermore, there is a promising avenue for advancement by introducing proppants with specific characteristics into fractures. These proppants not only facilitate better interactions but also contribute to maintaining fracture integrity. This innovative approach enhances foam stability and sustainability in fractured reservoirs, ultimately augmenting the entire foam generation process.

In the subsequent sections, we will delve into a comprehensive exploration of the fundamental tenets underpinning foam behavior, dissecting its intricate dynamics to offer an in-depth comprehension of its potential and application within porous media.

2.1 Foam definition

Foam is created when a gas phase is dispersed as small gas bubbles within a continuous liquid phase. Typically, the gas bubbles have diameters ranging from 10 to

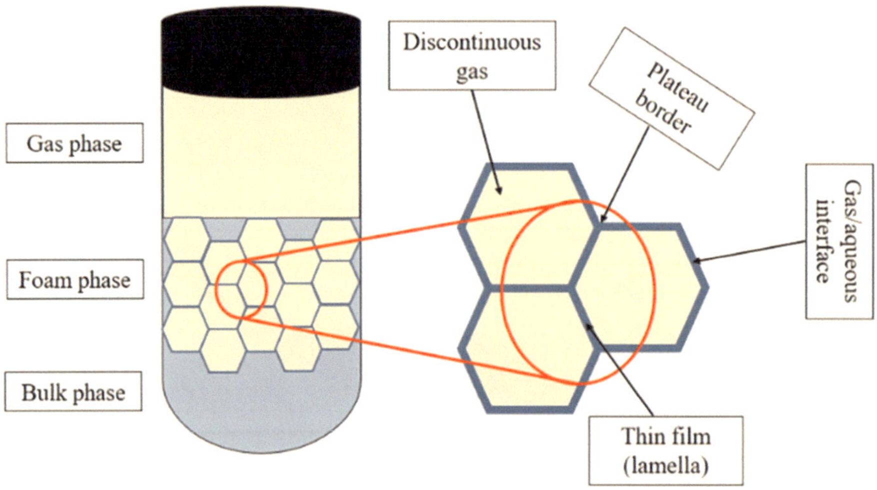

Figure 2.
A schematic of bubble generation.

1000 µm [22, 23]. The liquid phase, on the other hand, consists of surface-active materials, commonly known as surfactants. These surfactants are responsible for forming the interphase (i.e., thin film) between the gas and liquid called lamella as displayed in **Figure 2**. In other words, foam arises from the incorporation of gas bubbles into a liquid phase, with the liquid containing surfactants that stabilize the gas bubbles by forming an interfacial layer around them. This surfactant-laden interphase plays a pivotal role in creating and sustaining the foam structure. It ensures that the gas bubbles remain dispersed within the liquid phase, allowing foam to persist as a stable dispersion.

2.2 Foam texture

Foam texture refers to the physical appearance and structure of foam, which is a dispersion of gas bubbles in a liquid phase stabilized by surfactant molecules. It is impacted by the surfactant's composition and concentration, pore structure, and injection rates. Based on the bubble structure, foam can be distinguished as spherical and polyhedral foam as shown in **Figure 3** [24]. According to several previous investigations, spherical foam is a transient phase formed at the early stages of foam.

generation with thick lamella film, where foam is called wet foam with a higher proportion of liquid in comparison to gas content. However, polyhedral foam is a mature structure with thinner lamella film, and it is termed as dry foam as the water content is less than the gas content. On the other hand, based on the bubble size, foam is classified as coarse and fine [25]. In general, a foam with a coarse texture contains numerous large bubbles, also known as low-density foam, whereas a foam with a fine texture, referred to as high-density foam, is composed of a substantial number of small bubbles. Besides, coarse foam is generated at lower flow rates compared to fine foam generated at high flow rates. Several studies mentioned that spherical foam has smaller bubbles and more stable foam [26, 27] in contrast to polyhedral foam which possesses larger bubbles with less stability. Another opinion proposed that polyhedral foam is more stable compared to spherical foam [24].

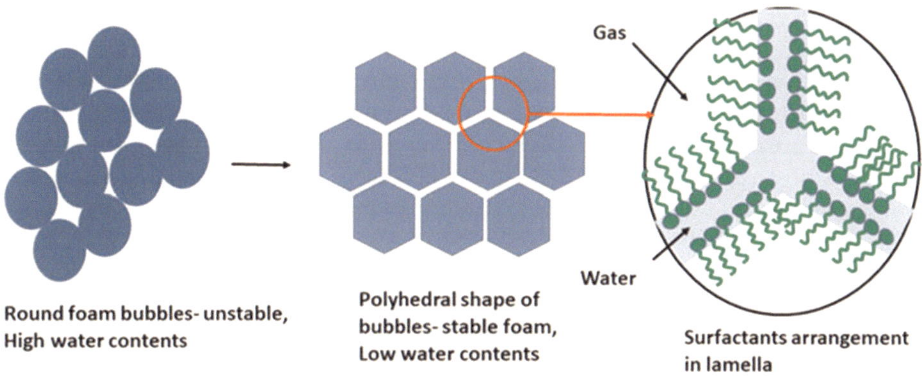

Figure 3.
Types of foam structures [24].

2.3 Foam generation mechanisms

Foam generation in porous media is accompanied by the creation number of lamellae films. These lamellae are created through three main mechanisms as demonstrated in **Figure 4**.

- Snap-off: It is the main mechanism for lamella creation in porous media. It occurs when a gas bubble becomes disconnected or snapped off from the continuous gas phase due to capillary forces or changes in pore geometry. As the gas bubble is disconnected, a thin liquid film wraps around it, creating a lamella. In fractured reservoirs, when gas is injected into the porous medium, it displaces the liquid phase (usually water or oil). As the gas advances, it can form elongated gas

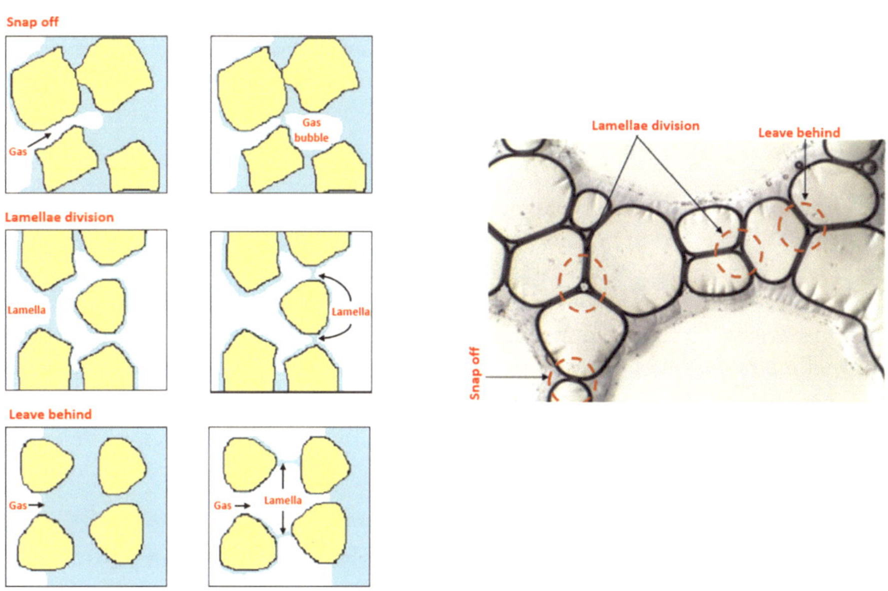

Figure 4.
Mechanisms of foam generation in porous media [28].

bubbles, and when certain conditions are met, these bubbles can become disconnected from the continuous gas phase. The process of snap-off occurs at pore throats or constrictions where the curvature of the gas–liquid interface is high. Capillary forces, which arise due to surface tension, act to minimize the surface area of the interface. When the curvature becomes too high, the capillary forces become dominant, and the gas bubble snaps off, leaving behind a liquid film that wraps around the bubble to form a lamella.

- Leave-Behind: In the leave-behind mechanism, as the gas bubbles move through the porous medium, they leave behind liquid films on the pore walls. These liquid films coalesce to form lamellae around the gas bubbles.

- Lamella Division: During foam flow through the porous medium, the lamellae can undergo division into smaller lamellae due to shear forces or interactions with the porous matrix. This process generates a higher number of lamellae and contributes to the stability of the foam.

Understanding the mechanisms behind lamella formation in porous media is essential for optimizing foam-based EOR strategies and achieving successful oil displacement and enhanced sweep efficiency. Researchers continue to explore these mechanisms to gain deeper insights into foam behavior and to develop more effective EOR techniques for various reservoir conditions.

3. Foam generation approaches

The selection of the most suitable foam generation technique depends on various factors, including reservoir properties, desired oil recovery goals, and operational considerations. Effective foam generation plays a crucial role in optimizing EOR processes and other applications that require efficient fluid displacement and mobility control. Several methods adopted by the industry and laboratory applications are summarized in **Figure 5** as follows:

3.1 Surfactant-alternating-gas (SAG)

This technique combines the advantages of surfactant injection with gas injection to create stable foams in the reservoir [29, 30]. Foam is formed inside the porous media by consecutive injection of gas and surfactant solution involving alternating cycles of surfactant injection and gas injection to create and maintain foam in the subsurface. This technique is alternatively known as "drainage foam" injection. Unlike traditional methods, foam generation using this approach is not confined solely to the entry zone. Instead, foam can be generated wherever there is contact between the gas and the invaded surfactant solution [30].

3.2 Foam pre-generation

This method involves the injection of a foam solution consisting of gas and a foaming agent into the oil reservoir. The foam is pre-formed before injection, which means it is generated at the surface or in a separate injection facility before being introduced into the reservoir [31]. The main findings of pre-generated foam injection

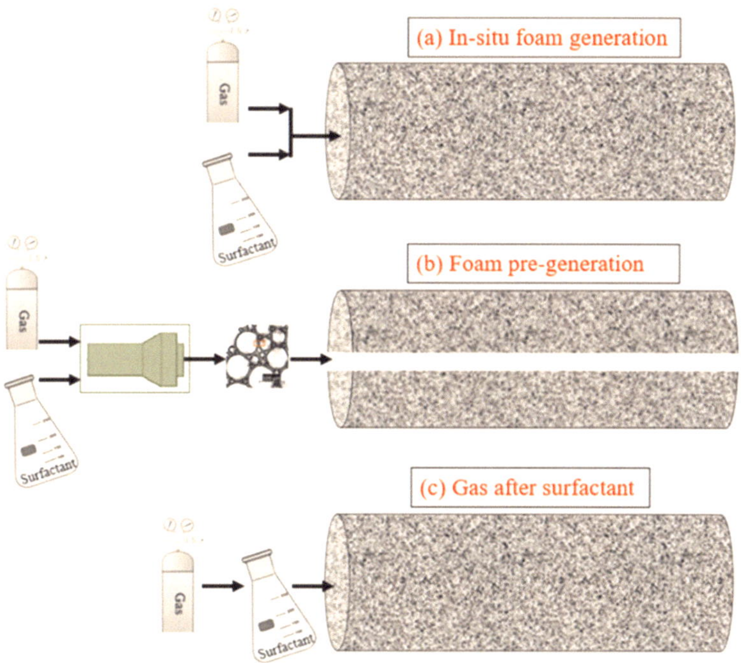

Figure 5.
Common foam generation methods.

for EOR include enhanced volumetric sweep efficiency, reduced gas mobility, and improved oil displacement, leading to higher oil recovery compared to conventional gas flooding methods, especially in fractured reservoirs. However, the success of this method depends on several factors such as reservoir properties, foaming agent selection, and injection parameters, which need to be carefully assessed and optimized for each reservoir to achieve the desired results. Furthermore, the properties of foam generated at the surface are different from foam in contact with porous media. It is crucial to ensure that the foam generated at the surface remains stable during injection, avoiding premature collapse or breaking down. As a result, there is a possibility of a gas breakthrough, where the gas phase separates from the liquid phase.

3.3 Co-injection

Co-injection method is another approach for foam generation. In this technique, both the gas and surfactant solution (foaming agent) are simultaneously injected into the reservoir through specific injection wells. The interaction of gas and surfactant at the subsurface conditions leads to the in-situ generation of foam, which aids in displacing and mobilizing the trapped oil [31]. During the co-injection foam process, monitoring techniques, such as tracer tests and pressure monitoring, are used to assess the performance and distribution of the foam in the reservoir. Adjustments in injection rates or surfactant concentrations may be made to optimize the process and maximize oil recovery.

This method offers several advantages, including increased oil recovery, better volumetric sweep efficiency, and the potential to mitigate gas mobility issues often encountered in conventional gas flooding processes. However, its applications in fractured reservoirs are limited and its successful implementation depends on understanding the reservoir characteristics and appropriately managing the injection parameters to achieve the best results.

3.4 In-situ co-injection in proppants

This technique resembles the co-injection foam process, where both the gas and the surfactant solution (foaming agent) are concurrently introduced into the reservoir. However, this approach is tailored specifically for fractured reservoirs to enhance the performance of foam. In essence, it involves the injection of proppants possessing specific characteristics into fractures, either naturally existing or created through hydraulic fracturing. This is done in conjunction with the simultaneous injection of foaming agents and gas, resulting in the generation of foam in situ [9, 32]. These proppants serve a dual purpose: they establish a porous medium that supports the formation of stable foam, and they adjust the fracture permeability for better fracture-matrix interactions. The strategic injection of proppants is particularly advantageous in the context of fractured reservoirs, encompassing diverse examples such as shale gas, tight oil, and coalbed methane reservoirs. However, thorough consideration and deliberate choice of the proppant variety, particle size distribution, and wetting characteristics emerge as critical factors to create more conducive conditions for effective foam generation. The selection of proppants should align with the specific reservoir's characteristics and objectives, aiming to establish an environment that optimally supports in-situ foam generation processes.

3.5 Dissolved surfactant in gas phase

Based on various research findings, certain surfactants have demonstrated the ability to dissolve in carbon dioxide when subjected to supercritical conditions [33, 34]. In this innovative approach, the injection process involves introducing a single phase into the reservoir, and the foam is generated as soon as this phase comes into contact with the formation water. This unique method takes advantage of the surfactant's solubility in supercritical carbon dioxide to simplify the injection process. Unlike traditional methods that require separate injections of gas and surfactant solutions, here, only one phase containing the surfactant is injected.

3.6 Simultaneous different layers

Several studies adopted this method to generate foam in the porous media where the gas phase and surfactant solution phase are injected simultaneously but in different layers of the well [35, 36]. In this method, the gas and surfactant solution are injected simultaneously, but in separate layers of the well. It is applicable to both vertical and horizontal wells. The gas injection occurs from the lower section or lower horizontal well, while the surfactant injection takes place from the upper section or upper horizontal well. The foam is generated inside the formation as a result of the natural separation caused by gravity, allowing the gas and surfactant phases to meet and interact.

4. Foam stability

Typically, foam stability is determined by the stability of lamellae film. The more stable the lamella film, the more stable the foam is, and vice versa. The stability of lamella film is controlled by several parameters such as pore geometry, experimental conditions, petrophysical properties, operating conditions, and surfactant formulation, and its concentration. These parameters shape the formation of bubbles, control gas diffusivity, disjoining pressure, and rate of capillary drainage.

4.1 Foam stability tests

4.1.1 Foamability

It measures how quickly foam generates by recording the pressure drop across porous media.

4.1.2 Foam half-life

The half-life of foam is the time it takes for the foam volume to reduce by half. This test assesses foam stability and longevity.

4.1.3 Foam apparent viscosity

It measures foam's ability to control gas mobility. The foam's apparent viscosity is typically determined using a rheometer, which measures the shear stress and shear rate applied to the foam sample. It also can be calculated using Darcy's law in a fractured system, as shown in Eq. (1).

$$\mu_f = K_{f-o} * \frac{\Delta P_{foam}}{\left(\frac{Q_g + Q_l}{A_f}\right) * L} \tag{1}$$

where K_{f-o} is the absolute permeability of the fracture to oil, A_f is the cross-sectional area of the fracture, Q_g, Q_l are the flow rates of the gas, and surfactant solution, respectively, and L is the length of the fracture.

4.1.4 Mobility reduction factor (MRF)

MRF is a critical parameter used to quantify the ability of foam to control gas mobility in porous media. It is calculated as the ratio of the gas mobility in the absence of foam to the gas mobility in the presence of foam as prescribed in Eq. (2).

$$MRF = \frac{\Delta P_{foam}}{\Delta P_{no\ foam}} \tag{2}$$

where ΔP_{foam} is the steady-state pressure drop across the porous medium due to foam generation and $\Delta P_{no\ foam}$ is the steady-state pressure drop without foam, i.e., co-injection of gas and brine (without surfactant).

4.2 Mechanisms of foam destabilization

Foam is destabilized by three underlying mechanisms including drainage, coarsening, and bubble coalescence. These mechanisms can lead to the breakdown of the foam structure and the reduction of foam stability:

- Coalescence: It occurs when two or more gas bubbles come into contact and merge to form a single, larger bubble. This phenomenon can happen when the liquid films separating the gas bubbles become too thin or rupture due to drainage or external forces.

- Coarsening: It is a gradual enlargement of gas bubbles in the foam over time. This occurs due to gas diffusion, where gas molecules migrate from smaller bubbles to larger ones. As smaller bubbles lose gas content, they shrink, while larger bubbles grow in size, leading to an increase in the average bubble size in the foam.

- Drainage: It includes the movement of liquid within the foam structure. As the foam sits or flows, the liquid films between gas bubbles can drain and become thin due to gravity or pressure gradients.

4.2.1 Gas diffusion

Gas diffusion refers to the movement of gas molecules within the liquid phase of the foam. This diffusion can have both positive and negative effects on foam stability. On one hand, a positive effect (stabilizing Effect) in which gas diffusion can contribute to foam stability by replenishing the gas content in the liquid phase. As the gas diffuses from the continuous gas phase to the lamellae and liquid films surrounding the bubbles, it helps maintain the gas content within the foam structure. This process helps sustain the foam's structure and prevents rapid bubble coalescence and foam collapse. On the other hand, a negative effect (gas Loss) where excessive gas diffusion can lead to gas loss from the foam structure. As gas molecules diffuse out of the foam into the surrounding liquid phase or porous matrix, the foam may lose its gas content, leading to the coalescence of the gas bubbles. This effect can be particularly pronounced in porous media with high gas permeability, where gas can easily escape from the foam structure.

4.2.2 Film drainage

In this mechanism, gravity causes the liquid inside the foam film to drain downward. Slowing down this drainage process can be achieved by increasing the bulk viscosity and reducing the liquid content of the foam. The time it takes for the lamella to reach critical thickness and the minimum thickness, leading to lamella coalescence, are crucial indicators of this process. These indicators depend on factors like surface elasticity, viscosity, gas-to-liquid ratio, surfactant solubility, and adsorption on the surface. Generally, the critical thickness diminishes as the surfactant concentration increases.

5. Factors affecting foam generation and stability

In fractured reservoirs, foam performance for EOR can be influenced by various parameters that impact foam generation, stability, and propagation through the fractures. Understanding these parameters is essential for optimizing foam-based EOR processes in such reservoirs. This can be achieved through laboratory experiments, reservoir simulation, and field trials resulting in a better understanding of the foam behavior and tailoring the EOR process to the specific reservoir conditions. Some of the key parameters affecting foam performance are considered below.

5.1 Fracture geometry

Fracture geometry plays a crucial role in influencing foaming ability, particularly in reservoirs with fractured formations. The geometry of fractures refers to their characteristics, such as aperture, orientation, connectivity, and distribution within the reservoir. These factors can significantly impact the behavior and efficiency of foam-based EOR techniques. **Table 1** lists all the factors and their impacts on foam performance due to fracture geometry.

By considering fracture geometry and its influence on foam transport in different directions, operators can tailor the injection strategy, foam composition, and operational parameters to achieve the best oil recovery results in such complex reservoirs.

5.2 Experimental conditions

This includes the pressure, temperature, and salinity of the reservoir. These conditions are critical factors that significantly influence foam performance in fractured reservoirs. They directly impact foam stability, propagation, and mobility control, which are vital for efficient oil displacement and improved recovery.

5.2.1 Operating pressure

Changes in reservoir pressure impact foam stability, foam generation, and gas solubility, ultimately affecting the efficiency of foam-based EOR techniques. In other words, at elevated pressures, the solubility of gas in the liquid phase decreases, leading to smaller gas bubbles and more stable foam. Stable foam can resist coalescence and collapse, ensuring its longevity and effectiveness in displacing oil. Furthermore,

#	Parameter	Impact
1	Aperture	The aperture of fractures, which refers to the width or opening between the fracture surfaces, influences the flow velocity and pressure gradients within the fractures. Narrower apertures may hinder foam propagation and distribution, while wider apertures can facilitate easier foam flow.
2	Orientation	The orientation of fractures in relation to the injection wells and the reservoir's geological features affects the direction of foam flow. Depending on the fracture orientation, foam may propagate more easily in certain directions, potentially leading to uneven oil displacement.
3	Fracture Connectivity	The connectivity of fractures determines how well they are interconnected within the reservoir. Well-connected fractures allow foam to propagate more effectively, covering a larger area and improving sweep efficiency.
4	Fracture Length and Width	Longer and wider fractures may influence foam transport and distribution. Longer fractures could lead to more extensive foam propagation, while wider fractures might accommodate more gas bubbles and stabilize the foam.

Table 1.
Factors arising from fracture geometry and their impact on foam performance.

operating pressure directly affects the generation of foam. For instance, at higher pressures, foam can be more effectively generated due to reduced gas solubility in the liquid phase. This can lead to more efficient foam propagation through the reservoir. It can also facilitate easier foam flow and distribution in fractures, enhancing the foam's contact with oil. In addition, higher pressure conditions often result in higher foam quality, where the gas phase occupies a larger fraction of the total foam volume. This can reduce gas mobility and delay gas breakthrough, leading to more uniform pressure fronts and better sweep efficiency. Last, high pressures can influence the capillary pressure of porous media and ensure the balance between foam stability and drainage.

5.2.2 Operating temperature

Reservoir temperature influences foam stability, gas solubility, viscosity, and phase behavior of the fluids within the reservoir, all of which can affect the effectiveness of foam performance in fractures. On one hand, the selection of a certain surfactant as a foaming agent is controlled by its susceptibility to temperature. It can impact surfactant adsorption at the gas–liquid interface and its ability to stabilize/destabilize the foam. Generally, lower temperatures favor foam stability as gas solubility decreases, leading to smaller gas bubbles and more stable foam. Higher temperatures can reduce foam stability by increasing gas solubility and promoting coalescence and bubble growth. In addition, changes in temperature can cause phase transitions in the reservoir fluids. For example, high temperatures may lead to gas condensation, altering the foam's properties and stability. On the other hand, temperature plays a crucial role in determining the foam rheology where foam viscosity increases as the temperature decreases, resulting in increasing the mobility reducing factor and providing better flow front to block gas path and therefore, diverting gas to inaccessible oil-filled zones.

It is essential to consider the impact of operating temperature when designing and implementing foam EOR processes, especially in unconventional reservoirs with fractures. Monitoring and managing reservoir temperature during EOR operations can help optimize foam performance and overall oil recovery. Reservoir simulations and laboratory experiments under various temperature conditions can provide valuable insights into foam behavior, guiding the selection of suitable surfactants, injection strategies, and operational parameters for successful foam EOR operations in different temperature environments.

5.2.3 Salinity

The formation salinity has a significant impact on foam performance and is a crucial factor to consider when selecting the appropriate surfactant for Enhanced Oil Recovery (EOR) processes. Salinity refers to the concentration of dissolved salts, such as Sodium chloride (NaCl), Calcium chloride ($CaCl_2$), Magnesium sulfate (MgSO4), Potassium chloride (KCl), etc. in the reservoir fluids. The salinity can vary widely from one reservoir to another and can fluctuate during production. The solubility of selected surfactants is affected by the salinity of the reservoir fluids. High salinity can lead to surfactant precipitation, reducing the amount of surfactant available to stabilize the foam. In extreme cases, surfactant precipitation may lead to the complete loss of foam stability. In addition, high salinity can increase the interfacial tension, making it more difficult for surfactants to reduce the tension and stabilize the foam.

Moreover, salinity can affect the micellar phase behavior of surfactants. In some cases, high salinity may lead to the breakdown of micelles, causing foam destabilization. Depending on the surfactant formulation, salinity could be a stabilizer or destabilizer of the generated foam. Accordingly, selecting the appropriate surfactant for foam applications in a specific reservoir requires considering the salinity conditions. Surfactants with good salinity tolerance are essential for stable foam performance in high-salinity reservoirs. Compatibility between the surfactant and the reservoir fluid's salinity is critical to ensure effective foam generation and stability. To this end, laboratory tests and reservoir simulation studies under various salinity conditions can help identify surfactants that are best suited for a particular reservoir's salinity range.

5.3 Surfactant formulation

The effectiveness of foam stability, particularly in challenging reservoir conditions, heavily relies on the presence of strong adsorption at the gas/liquid interface of foam lamellae. Foaming agents, such as surfactants, play a crucial role in achieving this goal, as they are capable of generating stable foams. These surfactants can be categorized into anionic, cationic, nonionic, and zwitterionic (amphoteric) families, based on their surface charge. The presence of surface charge on the surfactant molecules leads to a dense packing at the gas bubble interface, resulting in increased charge density and electrostatic repulsion, which prevents lamella film drainage. This closer packing also affects the shear viscosity, contributing to better foam stability [37].

Generally, ionic and amphoteric surfactants are frequently utilized as foaming agents due to the charged lamellae they form. The overlap of similarly charged electric double layers (EDL) prevents film thinning, thereby enhancing foamability and extending foam's half-life. On the other hand, nonionic surfactants lack electrical factors and are less effective at stabilizing foam [38].

Numerous studies have explored the impact of different surfactant solutions on foamability and foam stability. The findings regarding foamability are contradictory, with some studies favoring zwitterionic surfactants for better foamability, while others found nonionic surfactants to be more effective. One study even observed better foamability of anionic surfactants over zwitterionic ones [39]. In terms of stability, zwitterionic surfactants consistently exhibited higher foam half-lives compared to anionic surfactants.

The concentration of surfactant solutions also plays a significant role in foamability and foam stability. Studies have shown that foam properties depend on surfactant concentration and critical micelle concentration (CMC). Optimal foam properties are typically achieved at surfactant concentrations above CMC [40]. As the concentration increases, more surfactant molecules migrate to the liquid–gas interface, leading to a proportional increase in the initial foam height. Additionally, foam generated at concentrations above CMC exhibits a layering structure and reduced liquid film drainage. However, as surfactant concentration exceeds CMC, the foam's half-life dramatically decreases. The excess surfactant molecules inside the lamellae increase the gravitational effect of the foam, resulting in easier lamella drainage and eventual rupture of the foam film. Thus, careful consideration of the optimum surfactant concentration is essential to avoid foam collapse at high concentrations [41]. Balancing surfactant concentration is crucial for achieving stable and long-lasting foam performance in EOR applications.

5.4 Gas type

A variety of gasses such as nitrogen (N2), carbon dioxide (CO_2), hydrocarbon gases (methane (CH_4) and/or produced formation gases), and air, have been utilized in many investigations to generate foam [42–44]. It is noteworthy that some studies have suggested comparable mobility reduction factor (MRF) between foams generated using nitrogen and hydrocarbon gases like CH_4 [44]. However, other research has proposed that N_2 may perform better in porous media while being comparable to CO_2 and CH_4 in micromodels [43]. These discrepancies highlight the importance of considering various factors, such as reservoir properties, foam generation techniques, and the specific EOR application, when selecting the most suitable gas for foam generation.

Generally, the choice of gas for a particular application depends on a multitude of factors, each playing a significant role in determining the most suitable option. Some of these factors include the availability of the gas at a reasonable cost, the specific conditions of the reservoir, the type of recovery scheme being employed (whether miscible or immiscible), the properties of the recoverable oil, and the overall economics of the entire process, among others. When considering the selection of gas for foam-based, it is essential to assess the availability of the gas in sufficient quantities and at a reasonable price. Some gases may be readily accessible in certain regions, while others might be scarce or come at a higher cost due to transportation or production expenses. The practicality of using a specific gas will depend on its local availability and affordability. Moreover, reservoir conditions play a crucial role in determining the compatibility of gas with the specific geology and fluid dynamics of the reservoir. Different gases may interact differently with the reservoir rock and fluids, impacting the efficiency and effectiveness of the EOR process. Understanding the reservoir's unique characteristics is essential for choosing the most appropriate gas for foam generation or other EOR techniques.

5.5 Operating parameters

5.5.1 Injection rates

The injection rate influences the amount of gas that enters the porous media per unit of time. The higher the injection rates, the higher the pressure gradient, and the more gas enters the reservoir, increasing the potential for foam generation. Faster injection rates can lead to a higher gas fraction in the liquid phase, facilitating foam formation. However, at higher injection rates, there may not be enough time for foam drainage and film thinning between gas bubbles. As a result, gas bubbles may coalesce rapidly, leading to larger bubbles and destabilization of the foam structure. Furthermore, high rates can subject the foam to high shear forces, particularly at flow boundaries and around obstacles. These shear forces can disrupt the lamellae, which are the thin liquid films separating gas bubbles, leading to foam destabilization. Besides, extremely high injection rates can lead to large fluctuations in the pressure drop across the foam front. This sudden pressure drop can cause gas bubbles to expand rapidly, leading to foam instability and collapse. As a result, it must carefully select an optimal injection rate based on the specific reservoir conditions, foam formulation, and the desired objectives to avoid foam destabilization issued by injection rates.

5.5.2 Foam quality

The foam quality is a fraction of the foam volume containing gas. It can be defined when the relative amounts of gas and liquid are known. It is categorized into four regions based on the shape of foam bubbles, their interaction, and their effect on viscosity. In the first region (0–52% foam quality), spherical bubbles are uniformly dispersed without touching each other. In the second region (53–73% foam quality), the bubbles loosely arrange themselves in a cubical pattern, leading to increased viscosity during flow. The third region (74–95% foam quality) exhibits maximum foam viscosities, with bubbles forming parallel piped shapes. In the last region (above 95% foam quality), the foam is not stable as the liquid becomes the dispersed phase. Typically, foam regimes are classified into low-quality and high-quality regimes. In the low-quality regime, where the water fractional flow is high, bubble trapping and mobilization dominate the foam flow. As the gas fraction increases, the bubble volume increases, leading to drained lamellae films and collapse of the gas bubbles. This behavior occurs when the limiting capillary pressure (Pc^*) is reached and is better known as the dry-out effect when foam texture is dryer and coarser.

5.6 Petrophysical properties

Foam generation in porous media is influenced by various petrophysical properties of the rock. These properties impact how foam behaves as it flows through the rock's pore spaces, particularly in fractured reservoirs. Below are some of the properties that affect foam generation and its stability in porous media:

5.6.1 Rock wettability

Wettability is a fundamental physical property of reservoir rocks that has a significant impact on various parameters, including fluid distribution, relative permeabilities, capillary pressure, and residual oil saturation [27, 45]. It is influenced by factors such as rock mineralogy, fluid composition, saturation history, and reservoir temperature. Understanding wettability is crucial for predicting fluid behavior in reservoirs, optimizing production strategies, and designing enhanced oil recovery techniques. Wettability is proven to have a notable impact on foam generation and its stability in porous media. For instance, in water-wet rocks, where water has a higher affinity for the rock surface, foam generation is usually more favorable. This is because the presence of a water-wet surface facilitates the formation and stabilization of liquid films around the gas bubbles, promoting the stability of the foam structure. In oil-wet system, however, oil layer spreads over foam film resulting in collapsing lamella film. In addition, wettability impacts foam stability by influencing the drainage and coalescence of the gas bubbles. In water-wet rocks, the preferential wetting of the rock surface by the liquid phase (e.g., water) leads to slower drainage of liquid films between the gas bubbles. This reduced drainage can enhance foam stability as it prevents the gas bubbles from rapidly coalescing and merging, thus maintaining the foam structure for a longer time. Nevertheless, in oil-wet rocks, where oil has a higher affinity for the rock surface, foam generation and stability may be adversely affected. The presence of oil-wet surfaces can increase the resistance to foam flow, as the oil-wet rock may not provide the appropriate conditions for stable foam formation and propagation. This can lead to the breakdown of the foam structure and decreased foam stability.

5.6.2 Rock porosity and permeability

Fractured reservoirs are characterized by a complex network of interconnected fractures and matrix blocks, which introduce additional factors that influence foam behavior. Porosity and permeability have a double effect on foam performance. On one side, they affect the ease with which gas can flow through the porous media, and therefore, it directly impacts foam generation potential where higher permeability allows gas to flow more freely, making it easier to inject gas into the porous media and form gas bubbles, which are essential for foam formation. On the other side, porosity and permeability influence foam stability by affecting the drainage of liquid films between gas bubbles. In other words, in low-permeability media, the drainage of liquid films is slower, leading to more stable foam structures. This is because the liquid films have a longer residence time between gas bubbles, reducing the chances of rapid coalescence and bubble merging resulting in an improvement in the foam conformance and sweep efficiency during displacement processes. Accordingly, it is crucial to determine the optimum range of permeability that supports foam generation and also ensures the stabilization of gas bubbles.

5.6.3 Oil effect

The presence of oil has adverse effects on foam stability, making the detrimental impact prevailing. Surfactant interaction with oil causes liquid depletion in the thin liquid films (lamellae) that separate gas bubbles in the foam. Additionally, the presence of oil on the lamellae causes changes in rock wettability, and the spreading of oil disrupts the gas-water interface, leading to foam destabilization. Besides, the formation of an emulsion by the oil and surfactant solution further breaches the foam structure.

The mechanisms of foam stability against oil have been intensively studied, and Garrett proposed three coefficients to explain the oil's destabilizing effects on foam: the entering coefficient (E), the spreading coefficient (S), and the bridging coefficient (B). These coefficients evaluate the feasibility of oil droplets entering the gas-water surface and interacting with the foam phase as prescribed in Eqs. (3), (4), and (5). Positive values for E indicate easy penetration of oil into the foam, while positive values for S and B suggest oil spreading over the lamellae and acting as a bridge between gas bubbles, respectively as displayed in **Figure 6**. These actions lead to coalescence and reduced stability of the foam structure.

$$E = \sigma_{gw} + \sigma_{ow} - \sigma_{og} \tag{3}$$

$$S = \sigma_{gw} - \sigma_{ow} - \sigma_{og} \tag{4}$$

$$B = \sigma_{gw}^2 + \sigma_{ow}^2 - \sigma_{og}^2 \tag{5}$$

where, $\boldsymbol{\sigma_{gw}}$, $\boldsymbol{\sigma_{ow}}$, and $\boldsymbol{\sigma_{og}}$ are the surface tension between water and gas, interfacial tension between oil and water, and surface tension between oil and gas, respectively.

6. Conclusions and recommendations

Foam-based EOR can be a promising approach to enhance oil recovery and improve sweep efficiency in fractured systems. Fractured reservoirs present unique

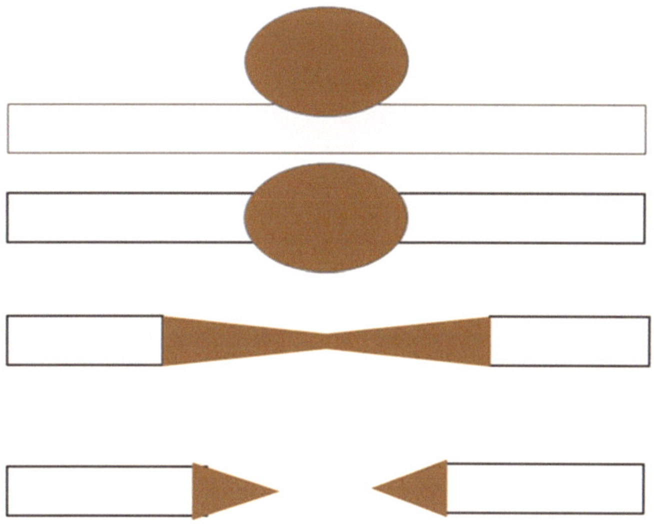

Figure 6.
Mechanisms of foam/oil interactions [24].

challenges due to their complex flow pathways, but foam-based processes offer several advantages in such environments. Foam can help address the mobility control issues caused by the presence of fractures, improve fluid distribution, and reduce the impact of channeling. Key findings and considerations for using foam EOR in fractured systems are:

- Mobility Control: Foam serves as a mobility control agent, reducing the mobility of gas and improving the sweep efficiency in fractured reservoirs. It can divert the injected gas into the matrix, where it displaces oil effectively, leading to better interactions between the fractures and matrix and therefore, enhanced oil recovery.

- Conformance Improvement: Foam can help mitigate the effects of permeability variations and fractures by improving conformance. It can divert gas into low-permeability zones and bypass high-permeability regions, and thus optimizing fluid displacement.

- Foam Stability: Foam stability is crucial in fractured reservoirs. The interaction between foam and the rock surface, along with the presence of oil, can impact foam stability. Optimizing foam formulations and injection parameters is essential to maintain foam stability throughout the injection process.

- Injection Strategy: The injection strategy, injection rates, foam quality, and experimental conditions play a significant role in foam performance in fractured systems. Properly designing the injection plan can help maximize foam sweep efficiency and minimize foam destabilization.

- Reservoir Characterization: Detailed reservoir characterization, including fracture mapping and permeability measurements, is essential to understand the fractured system's heterogeneity. This information guides the design of effective foam injection plans.

Acknowledgements

The author would like to acknowledge the University of Wyoming and the Center of Innovation for Flow through Porous Media for their support to gain knowledge and access to resources.

Conflict of interest

The author declares that they have no known competing financial interests or personal relationships that could have appeared to influence the work reported in this chapter.

Author details

Magda Ibrahim Youssif
University of Wyoming, Laramie, WY, USA

*Address all correspondence to: msolima1@uwyo.edu

References

[1] Liu G, Sorensen JA, Braunberger JR, Klenner R, Gorecki CD, Steadman EN. CO_2-based enhanced oil recovery from unconventional reservoirs: A case study of the Bakken formation. In: SPE Unconventional Resources Conference/ Gas Technology Symposium. The Woodlands, Texas, USA; 2014. DOI: 10.2118/168979-MS

[2] Kang WL, Zhou BB, Issakhov M, Gabdullin M. Advances in enhanced oil recovery technologies for low permeability reservoirs. Petroleum Science. 2022;**19**: 1622-1640. DOI: 10.1016/j.petsci. 2022.06.010

[3] Park H, Park Y, Lee Y, Sung W. Efficiency of enhanced oil recovery by injection of low-salinity water in barium-containing carbonate reservoirs. Petroleum Science. 2018;**15**:772-782. DOI: 10.1007/s12182-018-0244-z

[4] Liu P, Zhang X, Hao M, Liu J, Yuan Z. Parameter optimization of gas alternative water for CO_2 flooding in low permeability hydrocarbon reservoirs. Journal of Renewable and Sustainable Energy. 2016;**2016**:8. DOI: 10.1063/ 1.4948483

[5] Muther T, Qureshi HA, Syed FI, Aziz H, Siyal A, Dahaghi AK, et al. Unconventional hydrocarbon resources: Geological statistics, petrophysical characterization, and field development strategies. Journal of Petroleum Exploration and Production Technologies. 2022;**12**:1463-1488. DOI: 10.1007/s13202-021-01404-x

[6] Schön JH. Rocks their classification and general properties. Developments in Petroleum Science. 2015;**65**:1-19. DOI: 10.1016/B978-0-08-100404-3.00001-9

[7] Shojaei AK, Shao J. Application of continuum damage mechanics in hydraulic fracturing simulations. In: Porous Rock Fracture Mechanics: With Application to Hydraulic Fracturing, Drilling and Structural Engineering. Switzerland: Springer Nature; 2017. pp. 197-212. DOI: 10.1016/B978-0-08-100781-5.00009-9

[8] Gandossi L, Von Estorff U. An overview of hydraulic fracturing and other formation stimulation technologies for shale gas production. In: European Commission Joint Research Center Technical Reports. Luxembourg; 2013. p. 26347. DOI: 10.2790/379646

[9] Youssif MI, Sharma KV, Piri M. Hydrocarbon gas foam injection in fractured oil-wet carbonate samples: An experimental investigation of the effect of fracture-matrix permeability contrast on oil recovery. In: SPE Canadian Energy Technology Conference. Calgary, Alberta, Canada; 2023. DOI: 10.2118/ 212736-MS

[10] Le VS, Sharma KV, Hanamertani AS, Youssif MI, Elkhatib O, Rane K, et al. Methane foam performance in oil-wet unconsolidated porous media: A systematic experimental investigation at reservoir conditions. Fuel. 2023;**344**: 128002. DOI: 10.1016/j.fuel.2023.128002

[11] Wellington SL, Vinegar HJ. Surfactant-induced mobility control for carbon dioxide studied with computerized tomography. ACS. 1988; **1988**:344-358. DOI: 10.1021/bk-1988-0373.ch017

[12] Boerrigter PM, Verlaan ML, Yang D. EOR methods to enhance gas oil gravity drainage modeling aspects. In: Society of Petroleum Engineers - SPE/EAGE Reservoir Characterization and

Simulation Conference. Abu Dhabi, UAE; 2007. pp. 219-227. DOI: 10.3997/2214-4609-pdb.24.b20

[13] Wang L, Tian Y, Yu X, Wang C, Yao B, Wang S, et al. Advances in improved/enhanced oil recovery technologies for tight and shale reservoirs. Fuel. 2017;**210**:425-445. DOI: 10.1016/j.fuel.2017.08.095

[14] Rossen WR, van Duijn CJ. Gravity segregation in steady-state horizontal flow in homogeneous reservoirs. Journal of Petroleum Science and Engineering. 2004;**43**:99-111. DOI: 10.1016/j.petrol.2004.01.004

[15] Le Van S, Youssif MI, Hanamertani AS, Sharma KV, Elkhatib O, Rane K, et al. Methane foam performance evaluation in water-wet unconsolidated porous media: A systematic experimental investigation at elevated pressure and temperature conditions. Journal of Natural Gas Science and Engineering. 2022:104835. DOI: 10.1016/j.jngse.2022.104835

[16] Hanssen JE. Foam processes: An assessment of their potential in North Sea reservoirs based on a critical evaluation of current field experience. In: SPE Improved Oil Recovery Conference. Tulsa, Oklahoma, USA; 1994. DOI: 10.2118/27768-MS

[17] Zhu T, Ogbe DO, Khataniar S. Improving the foam performance for mobility control and improved sweep efficiency in gas flooding. Industrial and Engineering Chemistry Research. 2004;**43**:4413-4421. DOI: 10.1021/ie034021o

[18] Hirasaki GJ, Lawson JB. Mechanisms of foam flow in porous media: Apparent viscosity in smooth capillaries. Society of Petroleum Engineers Journal. 1985;**1985**:25. DOI: 10.2118/12129-PA

[19] Kovscek AR, Radke CJ. Fundamentals of foam transport in porous media. Advances in Chemistry. 1993;**1993**. DOI: 10.2172/10192736

[20] Farajzadeh R, Lotfollahi M, Eftekhari AA, Rossen WR, Hirasaki GJH. Effect of permeability on implicit-texture foam model parameters and the limiting capillary pressure. Energy and Fuels. 2015;**29**:3011-3018. DOI: 10.1021/acs.energyfuels.5b00248

[21] Rossen WR. Foams in enhanced oil recovery. In: Foams: Theory, Measurements, and Applications. 1996. DOI: 10.1201/9780203755709

[22] Gauglitz PA, Friedmann F, Kam SI, Rossen WR. Foam generation in porous media. In: SPE/DOE Improved Oil Recovery Symposium. Tulsa, Oklahoma; 2002. DOI: 10.2118/75177-MS

[23] Gauglitz PA, Friedmann F, Kam SI, Rossen WR. Foam generation in homogeneous porous media. Chemical Engineering Science. 2002;**57**:4037-4052. DOI: 10.1016/S0009-2509(02)00340-8

[24] Sagir M, Mushtaq M, Tahir MS, Tahir MB, Shaik AR. Foams in EOR. In: Surfactants for Enhanced Oil Recovery Applications. Switzerland: Springer International Publishing; 2020. pp. 41-63. DOI: 10.1007/978-3-030-18785-9_3

[25] Ibrahim KZ, Adebayo AR, Isah A, Muhammed NS. A literature review of strength and stability of foam and their relationship with the absolute permeability of porous media. Journal of Petroleum Science and Engineering. 2022;**2022**:211. DOI: 10.1016/j.petrol.2022.110195

[26] Audebert A, Beaufils S, Lechevalier V, Le Floch-Fouéré C, Saint-

Jalmes A, Pezennec S. How foam stability against drainage is affected by conditions of prior whey protein powder storage and dry-heating: A multidimensional experimental approach. Journal of Food Engineering. 2019;**242**:153-162. DOI: 10.1016/j.jfoodeng.2018.08.029

[27] Schramm LL, Mannhardt K. The effect of wettability on foam sensitivity to crude oil in porous media. Journal of Petroleum Science and Engineering. 1996;**15**(1): 101-113. DOI: 10.1016/0920-4105(95)00068-2

[28] Risal AR, Manan MA, Yekeen N, Azli NB, Samin AM, Tan XK. Experimental investigation of enhancement of carbon dioxide foam stability, pore plugging, and oil recovery in the presence of silica nanoparticles. Petroleum Science. 2019; **16**:344-356. DOI: 10.1007/s12182-018-0280-8

[29] Alcorn ZP, Fredriksen SB, Sharma M, Føyen T, Wergeland C, Fernø MA, et al. Core-scale sensitivity study of CO_2 foam injection strategies for mobility control, enhanced oil recovery, and CO_2 storage. In: E3S Web of Conferences. Vol. 146. France: EDP Sciences; 2020. DOI: 10.1051/e3sconf/202014602002

[30] Rossen WR, Boeije CS. Fitting foam simulation model parameters for SAG foam applications. In: SPE Enhanced Oil Recovery Conference. Kuala Lumpur, Malaysia; 2013. DOI: 10.2118/165282-MS

[31] Turta AT, Singhal AK. Field foam applications in enhanced oil recovery projects: Screening and design aspects. Journal of Canadian Petroleum Technology (Canada). 2002;**41**(10). DOI: 10.2118/02-10-14

[32] Youssif MI, Sharma KV, Piri M. Performance evaluation of innovative foam assisted gas injection schemes on propped-fractured oil-wet carbonate rocks at high-temperature and high-pressure conditions. Fuel. 2023;**2023**

[33] Le VQ, Nguyen QP. A novel foam concept with CO_2 dissolved surfactants. In: SPE Improved Oil Recovery Conference. Tulsa, Oklahoma, USA; 2008. DOI: 10.2118/113370-MS

[34] Ashoori E, Van Der Heijden TLM, Rossen WR. Fractional-flow theory of foam displacements with oil. SPE Journal. 2010;**15**(02):260-273. DOI: 10.2118/121579-PA

[35] Stone HL. A simultaneous water and gas flood design with extraordinary vertical gas sweep. In: SPE International Petroleum Conference in Mexico. Puebla Pue, Mexico; 2004. DOI: 10.2118/91724-MS

[36] Rossen WR, Vikingstad AK. Injection strategies to overcome gravity segregation in simultaneous gas and water injection into homogeneous reservoirs. SPE Journal. 2010;**15**(01): 76-90. DOI: 10.2118/99794-PA

[37] Clendennen SK, Boaz NW. Betaine amphoteric surfactants—Synthesis, properties, and applications. In: Biobased Surfactants. USA: AOCS Press; 2019. pp. 447-469. DOI: 10.1016/b978-0-12-812705-6.00014-9

[38] Chang Q. Emulsion, foam, and gel. In: Colloid and Interface Chemistry for Water Quality Control. London: Chemical Industry Press; 2016. pp. 227-245. DOI: 10.1016/B978-0-12-809315-3/00011-6

[39] Boeije CS, Rossen WR, Vad Bennetzen M. A methodology for screening surfactants for foam enhanced

oil recovery in an oil-wet reservoir. SPE Reservoir Evaluation and Engineering. 2017;**20**:795-808. DOI: 10.2118/185182-pa

[40] Lee J, Nikolov A, Wasan D. Foam stability: The importance of film size and the micellar structuring phenomenon. Canadian Journal of Chemical Engineering. 2014;**92**:2039-2045. DOI: 10.1002/cjce.22071

[41] Wang H, Chen J. A study on the permeability and flow behavior of surfactant foam in unconsolidated media. Environment and Earth Science. 2013;**68**:567-576. DOI: 10.1007/s12665-012-1760-6

[42] Andrianov A, Farajzadeh R, Mahmoodi NM, Talanana M, Zitha PLJ. Immiscible foam for enhancing oil recovery: Bulk and porous media experiments. Industrial and Engineering Chemistry Research. 2012;**51**:2214-2226. DOI: 10.1021/ie201872v

[43] Xu Z, Li B, Zhao H, He L, Liu Z, Chen D, et al. Investigation of the effect of nanoparticle-stabilized foam on EOR: Nitrogen foam and methane foam. ACS Omega. 2020;**5**:19092-19103. DOI: 10.1021/acsomega.0c02434

[44] Mannhardt K, Novosad JJ, Schramm LL. Foam/oil interactions at reservoir conditions. In: SPE/DOE Improved Oil Recovery Symposium. Tulsa, Oklahoma, USA; 1998. DOI: 10.2118/39681-MS

[45] Mannhardt K, Svorstøl I. Effect of oil saturation on foam propagation in snorre reservoir core. Journal of Petroleum Science and Engineering. 1999;**23**:189-200. DOI: 10.1016/S0920-4105(99)00016-9

Chapter 7

Advancements and Operational Insights in the Bakken Shale: An Integrated Analysis of Drilling, Completion, and Artificial Lift Practices

Ahmed Merzoug, Aimen Laalam, Lynn Helms, Habib Ouadi, John Harju and Olusegun Stanley Tomomewo

Abstract

This chapter provides an in-depth analysis of the Bakken Petroleum System (BPS) in the Williston Basin, focusing on Improved Oil Recovery (IOR) techniques. It explores the significant advancements in drilling, completion designs, and artificial lift methods that have markedly boosted oil recovery in this prime unconventional resource basin. The chapter traces the history of oil production in the Williston Basin, highlighting the transformative impact of horizontal drilling and multistage fracturing. It delves into advanced drilling operations, emphasizing the role of high-performance motors, geosteering, and real-time downhole data in enhancing drilling efficiency. Additionally, the chapter examines the evolution of well-completion strategies, from traditional to innovative horizontal completions, and assesses their effectiveness through data analytics, numerical modeling, and field studies. The vital role of artificial lift systems in combating rapid production decline in shale formations is analyzed, comparing the efficacy of ESPs, Sucker Rod Pumps, and Gas Lifts. The interconnectivity between operational aspects is discussed, providing a unified view of how integrated strategies and technological advancements drive optimized oil recovery in the Bakken formation. This study aims to offer insights and strategic guidance for industry stakeholders, particularly concerning IOR in unconventional oil resources.

Keywords: Bakken shale, drilling practices, completion design, artificial lift optimization, unconventional oil production

1. Introduction

The Williston Basin holds the distinction of being among the earliest consistent producers of shale oil resources in the United States. Oil was first discovered in the basin in 1936, and the region became a major oil province in the 1950s with the discovery of large fields in North Dakota. Production initially peaked in 1986, but significant increases in production began in the early 2000s due to the application of horizontal drilling techniques, especially in the Bakken Formation [1]. A myriad of successful drilling and completion methodologies took root and were developed within the context of the Williston Basin, and these have since been adopted and implemented nationwide. In its nascent production phase, it constituted the highest production yield from unconventional resource basins in the United States. As of June 2023, the Williston Basin continues to be a significant player in the United States' oil production landscape, ranking as the second-largest producing region in the country, surpassed only by the Permian Basin [2].

In 2010, the Energy & Environmental Research Center (EERC) executed an in-depth, multidisciplinary research program to rigorously examine the key elements influencing successful oil production in the BPS, North Dakota. This study emphasized four primary areas: geology, geochemistry, geomechanics, and drilling/completion engineering. Initial results underscored the efficacy of horizontal drilling in conjunction with multistage fracturing, the role of geological variables in determining hydrocarbon production rates, and increased production in Mountrail County, linked to heightened concentrations of organic carbon and thicker shale deposits. Emphasizing optimal completion practices involving horizontal drilling and fracture stimulation is fundamental to liberating reservoir fluids held tightly within the formation. Notably, horizontal drilling of the Bakken's middle member combined with multistage fracturing has surpassed the performance of all prior Bakken wells in North Dakota. On the contrary, multilateral wells, despite their reduced per-foot drilling costs, seem to present no significant advantage in terms of production. The study revealed that in the Williston basin, wells in Mountrail County, one of the largest producers, predominantly used heavy, oil-based mud for drilling, whereas Dunn County wells employed water-based brine. These drilling fluids can cause harm to horizontal wellbores and impact their wettability. Although water-based fluids facilitate quicker and more economical drilling, they lead to increased pipe wear and come with density limitations. Conversely, oil-based fluids lessen drag and pipe wear and provide more flexibility in density. Oil-based drilling muds are crucial for the installation of oil-based swell packers used in multistage fracturing [3].

Based on data from May 2023, the top producing operators in North Dakota in terms of oil and gas production have shown some changes. Continental Resources, Inc. remains a dominant player, producing 6,254,994 barrels of oil and 18,699,327 MCF of gas from 2179 active wells. They're followed by Marathon Oil Company and Hess Bakken Investments II, LLC, which produced 2,550,434 and 2,298,305 barrels of oil, respectively. Burlington Resources Oil & Gas Company LP has made a notable jump in its contributions, producing 1,991,140 barrels of oil. On the gas production side, Hess Bakken Investments II, LLC remains a significant contributor with 9,638,090 MCF, while Whiting Oil And Gas Corporation produced 7,409,455 MCF. In terms of active wells, Continental Resources, Inc., Whiting Oil And Gas Corporation, and Hess Bakken Investments II, LLC continue to be major operators with 2179, 1786, and 1423 wells respectively. The wells with the highest oil production have shifted. The 'GT USA 11-18H' operated by Marathon Oil Company in Dunn County stands out with 65,539 barrels of oil. Following closely are the 'Dennis Fiu' series of wells, particularly 'Dennis Fiu 4-8H' operated by Continental Resources, Inc. in Dunn County producing 52,828 barrels. As for gas production, the 'Sorenson Federal 153-96-9-4-14H' well operated by Ovintiv USA Inc. in McKenzie County dominates, generating a whopping 146,161 MCF of gas. Interestingly, many of the top producing wells are located in Dunn County, with Continental Resources, Inc. being a particularly dominant operator in the region [4].

The drilling statistics report from the North Dakota Industrial Commission from 1951 to 2021 indicates a substantial growth in total footage over the years, peaking in 2014 at 44,941,118. However, a decline occurred afterward, likely attributed to the impact of the 2015 oil price collapse followed by the COVID-19 pandemic. Nevertheless, there was a slight recovery observed in 2021, with the total footage reaching 14,129,194 compared to 12,914,547 ft. in 2020. These statistics highlight the consistent drilling activity in North Dakota over the years, resulting in a combined footage of 384,882,036 [5].

Figure 1 illustrates the initial oil production, initial water cut (WC), initial Gas Oil Ratio (GOR), and wells drilled per year for the BPS. The average values per year are reported. The graph shows a steady increase in initial oil production (24 h), an increase in initial GOR and initial water cut. These observations are attributed to the development of several technologies that permitted the increase in the performance of these wells. The initial water cut trend is a result of an increase in hydraulic fracturing treatment volumes described by Miller et al. [6]. The increase in GOR trend is attributed to the tighter well and fracture spacing explained by Acuña [7]. The implemented fracture spacing results in smaller rock fragments, thus less pressure support. This will cause the pressure in these fragments to decrease rapidly below bubble point pressure resulting in higher GOR. When offset wells are drilled next to existing wells the rock can already be below bubble point pressure resulting in higher GOR. We note the increase in the number of child wells reported by Latrach et al. [8]. The well GOR per location during different periods is illustrated in **Figure 2**, the figure shows a relatively low initial GOR at early development (2005–2010). After this period, well density increased at the center of the basin with higher GOR. Wells that are further from the highest well density area had a lower initial GOR (**Figure 3**).

The performance change in **Figure 2** is not representative of the whole basin as there some are companies that perform better than others in terms of initial production. In this study, only the top 6 operators are studied. The top 6 are selected based on the number of wells they own in the Bakken. **Figure 4** illustrates the evolution of initial production parameters for these operators. Even though all operators showed an increase in initial production over time, they do not perform the same. **Figure 5** illustrates the distribution of well locations per operator. Even though operators have wells within reasonable proximity of one another. These wells did not perform the same; thus the well's initial production is likely a result of operators' approach for development.

This research endeavor represents an ambitious attempt to unravel the multiple changing variables that characterize oil production within the BPS. It aims to shed

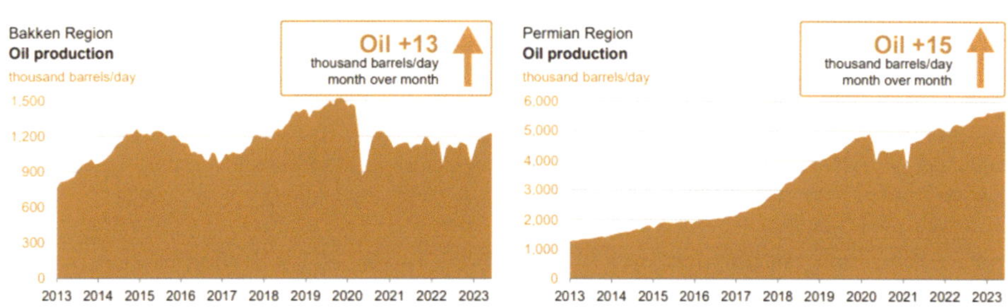

Figure 1.
Oil production in the Bakken and Permian regions (in thousands of barrels per day). Data source: U.S. Energy Information Administration [2].

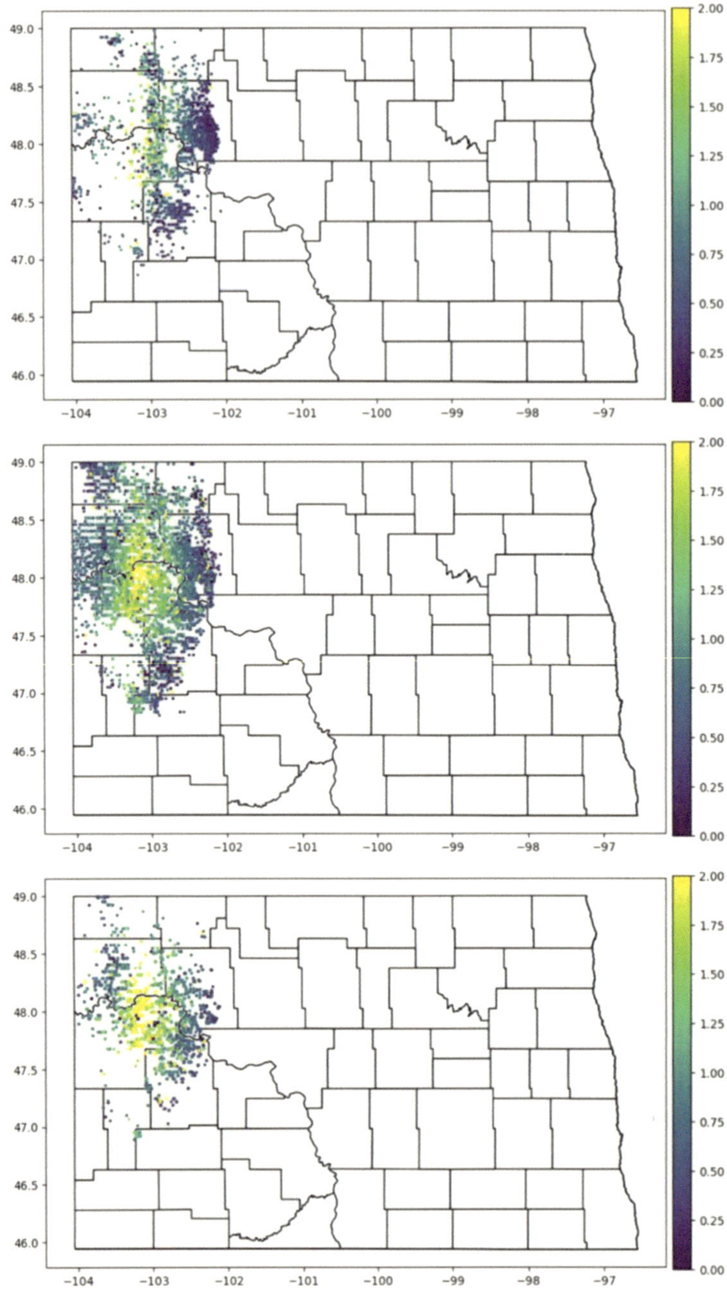

Figure 2.
Initial GOR per well location. (a) Wells drilled between 2005 and 2010, (b) wells drilled between 2010 and 2015, (c) wells drilled between 2015 and 2023.

light on the divergent performance of different companies, track the evolution of their outputs over time, and contextualize these within the broader trajectory of oil production within the BPS. The primary focus of this paper is to perform a thorough literature review and data analytics study to interpret observable trends in three key areas: drilling practices, artificial lift mechanisms, and completion design.

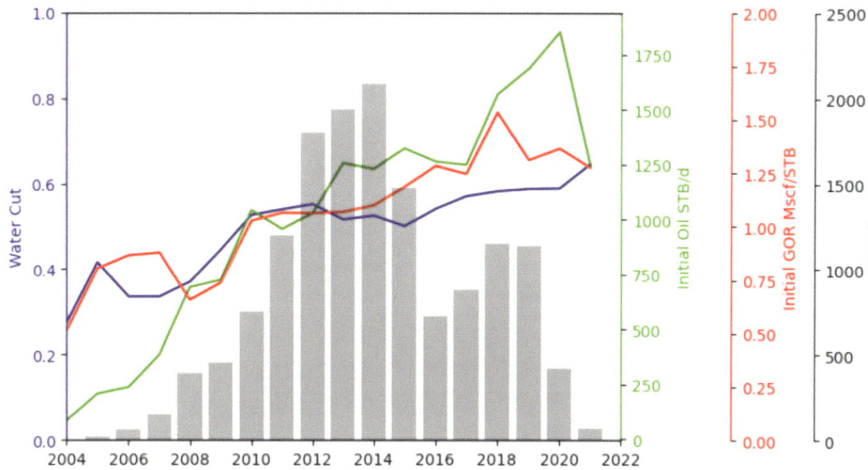

Figure 3.
Average initial oil production, initial GOR, initial water cut, and wells count in the BPS.

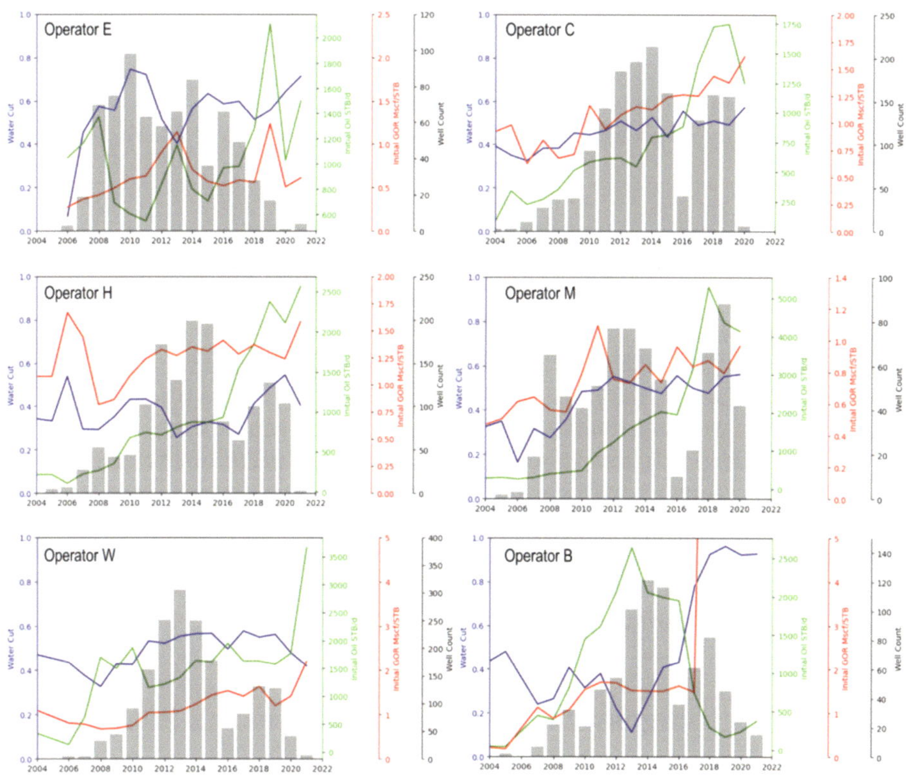

Figure 4.
Average initial oil production, initial GOR, initial water cut and wells count per operator in the BPS.

The methodology of this study is twofold. Firstly, the performance of each operator is meticulously examined, thereby enabling a detailed comparison between the players in the field. Our focus is particularly geared towards those producers who have demonstrated a consistent increase in production and have an extensive array of wells

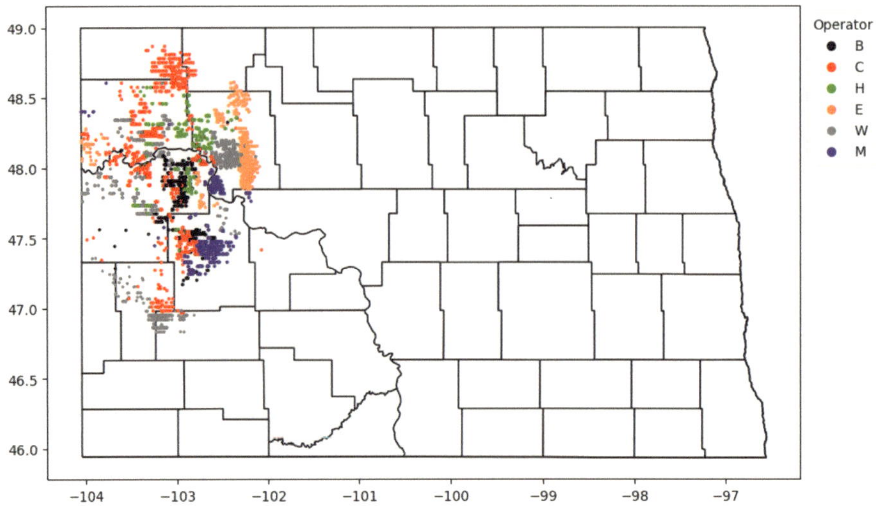

Figure 5.
Wells location per operator.

that underpin their operational success. Secondly, through a comprehensive review of academic articles and industry publications associated with each operator, we gain insights into their specific practices, technological preferences, and operational strategies.

2. Key insights into the Bakken formation's role in oil production

The Bakken Formation, spanning across parts of North Dakota, Montana, Saskatchewan, and Manitoba, is a significant oil and natural gas reserve in the United States [9]. Dating back to the Late Devonian to Early Mississippian periods, this formation within the Williston Basin is characterized by layers of shale, siltstone, and sandstone [10]. Originally considered marginal due to low permeability, technological advancements in horizontal drilling and hydrofracturing have transformed it into a major oil and natural gas producer [11]. By 2012, these developments elevated North Dakota to the second-largest oil producer in the U.S., behind Texas [12]. The USGS estimates substantial undiscovered resources in the Bakken Formation, making it a key player in North America's energy sector [13].

3. Exploring the background of oil and gas drilling in the Williston Basin

3.1 General well design and challenges in the bakken formation

The Djurisic et al. [14] study describes a three-phase well drilling design in the Bakken region:

- Surface Interval: A 13–1/2 inch hole is drilled to 2100 feet using a rock bit. It involves cementing 9–5/8 inch casings with a rotary assembly. Challenges include managing winter water and maintaining hole inclination.

DOI: http://dx.doi.org/10.5772/intechopen.1003955

- Intermediate Interval: Drilling reaches the kickoff point with 8–3/4 inch PDC bits, then curves towards the Middle Bakken horizon. The section uses oil-based mud and a 7-inch casing. Main challenges are PDC bit wear and geological considerations.

- Production Interval: A lateral hole is drilled to about 20,000 feet with 6-inch PDC bits. The stage uses brine mud and possibly a 4–1/2-inch liner. Challenges here include steering accuracy, managing BHA walk, torque, tool reliability, and handling casing wear and well control.

3.2 Drilling operations and technologies

In North Dakota's horizontal drilling operations, conventional mud motors and Measurement While Drilling/Logging While Drilling (MWD/LWD) systems, along with onsite geological analysis, are frequently used for drilling curve and lateral sections. Positive displacement mud motors with Adjustable Kick-Off angles are used in drilling the curve. The MWD system measures the earth's gravity and magnetic field, and wellbore inclination with survey stations crucial for tracking the wellbore's position. These systems need sufficient non-magnetic spacing and proper magnetic corrections for accuracy. The temperature sensor is another vital component, as most sensors are temperature calibrated. The placement of LWD gamma ray modules varies across service providers [15].

3.3 Key studies and their findings

This collection of research papers provides a detailed insight into the advancements in drilling technologies and efficiencies in the Bakken and Williston Basin Petroleum Systems, focusing on several key technical aspects:

Djurisic et al. [14] demonstrated a significant 50% increase in the Rate of Penetration (ROP) over 18 months, attributing this improvement to advancements in drilling technology and techniques, including the use of high-performance motors with pre-contoured stators and real-time downhole drilling data analysis. This enabled a more profound understanding of critical drilling parameters such as weight on bit and torque. Pearson et al. [16] discussed how technological advances have led to a substantial boost in oil production, even amidst falling oil prices and economic challenges. They highlighted how strategic changes in completion performance and well design, such as transitioning from Generation 1 to Generation 4 and then Generation 5 designs, significantly improved drilling returns. Lolon [17] focused on the success of horizontal well completions in the Bakken formation, noting a threefold increase in production rates compared to vertical wells. The study emphasized the effectiveness of longer laterals, staged treatments, and cleaner fluids, along with the importance of fracture design and proppant selection. Johnson and Courrege [18] reported on the increased oil and gas production in the Bakken Shale using openhole packer and sleeve completions. This technology proved beneficial in maximizing exposure to hydrocarbon reserves and reducing environmental impacts.

Southcott and Harper [19] detailed WPX Energy's significant reduction in geosteering errors through the advanced processing of 3D seismic data. This technological advancement led to more accurate mapping and lower well costs. Brandt et al. [20] presented an engineering-based evaluation of the energy intensity and net energy return (NER) in oil production. They highlighted the increasing energy cost associated with drilling, underlining a declining trend in NERs. Gutierrez et al. [21] introduced a geosteering methodology using azimuthally sensitive gamma-ray sensors for precise well positioning, leading to enhanced production rates.

Lim et al. [22] emphasized the integration of production engineering with drilling strategies, using Torque & Drag (T&D) modeling to ensure quality wellbores and facilitate successful liner installation. Veazey [23] reported on Whiting Petroleum Corp.'s strategic expansion in the Williston Basin to deepen its drilling inventory, while Halliburton [24] introduced the Cerebro intelligent bit technology, enhancing drilling efficiency through high-frequency motion mapping and vibration measurement. Jerrard et al. [25] stressed the importance of real-time observation in cement operations, suggesting adjustments in slurry design based on bottom hole temperature observations. Halliburton [26] discussed a customized water-based fluid system for drilling, focusing on wellbore stability and contamination prevention, while Halliburton [27] introduced NitroForce®, a high-torque motor technology that significantly reduced well time and increased drilling efficiency. Gundersen et al. [28] explored the implementation of Wet Shoetrack Completions, a cost-efficient technique for unconventional oil and gas extraction, and Schmidt et al. [29] examined well-to-well interactions, suggesting that sacrificial fracs in child wells could optimize production. Lastly, Ouadi et al. [30, 31] introduced the innovative fishbone drilling method, which increases hydrocarbon recovery and reduces environmental impact by featuring a central wellbore with multiple branching laterals.

3.4 Overcoming obstacles in North Dakota's drilling industry

Reddy and Pitcher's [32] case study highlighted the complexities encountered in geosteering a horizontal well in the Middle Bakken Dolomite. The operator initially faced difficulties with the wellbore straying from the desired zone, necessitating multiple open-hole sidetracks. To mitigate these challenges, they deployed a comprehensive geosteering service that included an azimuthal resistivity tool, dedicated software, and specialist personnel, leading to more accurate well positioning and reduced need for costly unplanned sidetracks. Bassarath and Maranuk's [33] research focused on the challenges of drilling long lateral wells in the Bakken shale, introducing the Targeted Bit Speed (TBS) technology. This technology significantly improved 3D directional control, akin to a rotary steerable system, and was instrumental in reducing sliding time, increasing the Rate of Penetration (ROP), and drilling smoother wellbores, thus enhancing the overall efficiency of shale drilling operations.

McCormick et al.'s [34] study proposed the use of expandable liner hangers (ELHs) for the completion of long horizontal wells in shale plays, notably in the Bakken Shale. ELH systems, which combine the liner hanger, packer, and tieback receptacle into one unit, simplify operations and have demonstrated excellent performance in managing large loads, thereby increasing reliability, safety, and profitability in shale operations. The study by Parayno et al. [35] addressed the drilling challenges in the Bakken shale and suggested Managed Pressure Drilling (MPD) as an effective solution. MPD, which allows for instant changes to wellbore pressure without regular mud weight adjustments, led to improved drilling efficiency, better hole cleaning, and fewer issues related to downhole pressures. Chinander's [36] study, along with Tobben's [37] report, discussed the fluctuating state of oil production in the Bakken region, peaking in late 2019 and then declining in 2020 due to the COVID-19 crisis. These studies raised concerns about the future productivity of the Bakken Shale formation, especially given the increased gas production from mature wells, which is negatively affecting crude output. Additionally, the Western Dakota Energy Association [38] highlighted the potential advantages of extending underground drilling laterals from two miles to three miles, particularly in less productive areas of the Bakken. This approach could lead to fewer overall wells needed and a shift towards enhanced oil recovery in more extensively drilled areas. Lastly, Jean's [39] article in the Williston Herald reported on North Dakota's recent peak in rig counts, which is being stifled by severe labor shortages and infrastructure limitations. Despite rising oil prices, the state faces challenges in expanding its labor pool and accommodating the increasing gas-to-oil ratio with adequate infrastructure.

4. Hydraulic fracturing and completion design

This section attempts to summarize the most relevant research and publications related to hydraulic fracturing and completion design in the Bakken Petroleum System. It tracks the latest technologies and their applications. The section is organized into three subsections.

The first subsection is statistical studies. It summarizes the statistical analysis conducted by top operators to understand production efficiency related to different completion designs. It also reports some statistical findings related to refracturing, parent-child wells interference, drilling directions, and stress estimation using Diagnostic Fracture Injection Tests (DFITs).

The second subsection is field pilot studies. It summarizes testing that was conducted in the Bakken. Operators tried to understand the effect of completion design in the Bakken Petroleum System by testing different ideas in the field. In their attempt they used several diagnostic approaches and techniques to understand the completion design and technologies. They also report some of the challenges and how they overcame them.

The third and last subsection is numerical modeling studies. This subsection reports modeling efforts that were conducted to history match and pilot field studies. These efforts leverage physics-based modeling to understand the behavior of the reservoir and draw conclusions about the efficiency of hydraulic fracturing and completion design in the Bakken.

4.1 Statistical studies

Griffin et al. [40] conducted a comparative statistical analysis of the performance of different operators in the Bakken. They found out that operators with advanced completions (**Table 1**) performed better than their competitors in terms of averaged cumulative production and EUR. The performance is notably different in higher water cut wells. The authors state that the use of ceramic proppant results in better production; however, we believe that the uplift of production is related to the use of higher injection volumes because, comparing operators B and E, the injection and proppant volumes are the same; however, the proppant types were different with higher proppant cost for operator E and lower EUR and cumulative production. They also noted that cumulative water cut represented a very good proxy for reservoir quality (lower water cut indicates better reservoir quality). Their study was limited to a well basis, and not to a drilling and spacing unit (DSU). Note that in unconventional plays development, the fractures from these wells create a connected system that imposes an interdependency of well performance.

Operator	Wells #	Liner	Stages #	Comp. Type	Fluid	Prop. Lbs/ ft	Vol. Bbls/ ft	Sand %	Ceramic %	Cost MM$	year Cum	EUR MBOE
A&C	45	SP	35	PNP	SW	396	25.1	0	100	4.2	195	1050
B	144	SP	34	PNP	Hyb.	395	7.9	38	62	3	140	800
E	56	Cmt	25	PNP	Gel	353	7.8	0	100	3	124	700
L	157	SP	30	PNP	Gel	288	5.7	68	31	2.3	105	600
Y&T	68	SP	28	BS	Gel	300	6.1	83	15	2.3	107	600
R	76	SP	25	PNP	Gel	264	6.5	100	0	1.9	87	500

Table 1.
Averaged values for completion properties and performance for different operators for 30% water cut [40].

Lolon et al. [41] built a multivariate statistical machine-learning model. To evaluate the impact of different fracture designs on well performance. They used data from 6800 wells from the North Dakota Industrial Commission and Frac Focus. They used root mean square error in cross-validation to measure the performance of their models. They reported that water-cut is inversely proportional to cumulative oil production. They also noted that water-cut can be used as a proxy for reservoir quality. The TVD was a proxy for oil maturity. Higher treatment rates, volume and proppant mass had a positive effect on production (**Figure 6**). Note that these effects are not absolute as multiple parameters change from one geographic area to another one. The paper outline sensitivity analysis for these different parameters using the machine learning models to quantify the effect of changing one parameter at a time to support the previous conclusions.

Figure 6 illustrates the Pearson Coefficient of Correlation for various factors impacting oil recovery in the Bakken Petroleum System, revealing both positive and negative relationships. Higher water cuts and increased stage spacing show negative correlations, suggesting that excessive water production can impede oil flow and wider fracturing stages might lead to a less efficient fracture network, respectively. This indicates that optimizing water management and stage spacing could be crucial for enhancing oil recovery. In contrast, a positive correlation with treatment volume and proppant mass emphasizes the benefits of sufficient hydraulic fracturing inputs in creating expansive, conductive fracture networks that improve hydrocarbon migration. Interestingly, the organic content as measured by TOC shows a contrasting effect; while higher TOC in the upper Bakken correlates negatively with oil recovery, suggesting complexities in geological variability or non-productive organic matter, the positive correlation with TOC in the lower Bakken implies a region richer in hydrocarbon-generating organic matter. The analysis underscores the nuanced

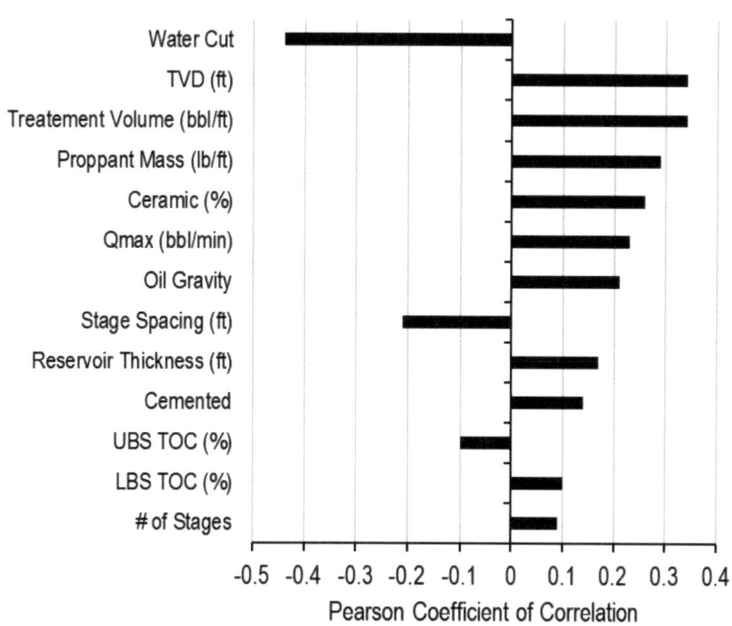

Figure 6.
Correlation between reservoir and completion parameters to 180 days oil cumulative production normalized per lateral length modified from [41].

interplay between geological, operational, and technical factors in optimizing oil recovery in unconventional shale plays.

Pearson et al. [42] emphasized the importance of near-wellbore conductivity. They supported their idea with the fact that most of the pressure drop occurs near the wellbore; thus higher proppant conductivity is required. They ran a series of laboratory tests and simulations in the BPS to support their idea. The tests showed a time-dependent proppant conductivity under the same stress condition. They conducted a numerical simulation to history match well production in the Bakken. The model was then used to sensitize on using different proppant combinations of ceramic and brown sand with ratios varying from 0 to 100% of the total injected proppant. The time of injecting ceramic proppant was also changed between lead and tail for different concentrations. The economics were also evaluated for these cases showing a significant incremental for using a 5% lead, 5% tail ceramic proppant. The latter design only increases the completion cost by 8% but results in an incremental oil recovery of 8%. Their study showed the potential production uplift of having higher conductivity in the near wellbore region.

Taghavinejad et al. [43] conducted a Rate Transient Analysis study on 73 wells of the Bakken and Three Forks Formations to quantify the effect of Fracture Driven Interaction (i.e., frac-hit). Their work divided the BPS wells into 3 generations, summarized in **Table 2**. Note that the pump rates contradict other papers, especially for Gen 2 and Gen 3 actual rates and sand types. In their work, they concluded that the Fracture Driven Interaction (FDI) is distance and production time-dependent. FDI resulted in 50% losses in EUR, up to 80% incremental water for first Gen wells. Increased proppant and fluid volume increased production only up to a certain value above which the increased volumes did not have a significant effect. These results can be area specific where other responses have been noted in the following studies.

Cozby and Sharma [44] conducted a statistical study on the impact of FDI on the production of parent and child wells. Their study covered several plays across the United States, including the Bakken play. Previous to their study, there were several papers written on the same topic, including [6, 45]. The results from these studies showed that on average, BPS' offset wells overperform BPS' parent wells. They also found that wells

Metric	**Gen 1 (Parent)**	**Gen 2 (Infill)**	**Gen 3 (Infill)**
Completion Date	2009–2012	2012–2015	2015–2016
Formation	Three Forks (TF)	70% TF, 30% Bakken	45% TF, 55% Bakken
Completion Type	OH SS	OH SS, PNP	PNP
Number of Stages	10–26	17–30	26–30
Cluster per stage	1	1–5	5
Stage Spacing (ft)	390–970	319–593	310–460
Sand Concentration (lbs/ft)	85–225	160–625	195–280
Sand Type	20/40 and 20/70	20/40 and 20/70	30/30 and 40/70
Fluid Volume (bbl/ft)	3–5	4–9	4–8
Fluid Type	Slickwater/ XL tail-in	Slickwater/ XL tail-in	Slickwater/ XL tail-in
Pump Rate (BMP)	46	46	46

Table 2.
Evolution of hydraulic fracture job design in the BPS [43].

in the BPS that had been drilled and stimulated before 2017 resulted in incremental production due to FDI, whereas wells that were drilled after 2017 resulted in underperformance in the parent wells. This was attributed to the evolution of completion size. Wells drilled before 2017 had a smaller completion size resulting in an under-stimulated volume of the reservoir accessed once the child well was drilled and fractured. They reported that the two main factors that affect the parent well response are well spacing and the Cumulative production volume before the stimulation of the child well (**Figure 7**). The reader is referred to Gupta et al. [46] for a complete review on FDI.

The heatmap presented in the **Figure 7** provides a visual representation of the relationship between the volume of Barrels of Oil Equivalent (BOE) produced by parent wells and the spacing between drilling sites, along with the corresponding change in production over a 12-month period. The left panel of the heatmap indicates that as the spacing between wells increases, there is a trend towards greater initial production volumes, denoted by the transition from cooler to warmer colors. This suggests that wider well spacing may lead to less competition for resources between wells and potentially enhance individual well performance. The right panel, which shows the data count, reflects the distribution of data points across different spacings and production changes, indicating the most commonly encountered scenarios. Together, these heatmaps provide a dual perspective, linking well spacing to production efficacy and highlighting areas where data are most abundant, thereby offering strategic insights for drilling and development planning in the context of Improved Oil Recovery (IOR) efforts in shale formations.

Other than FDI, refracturing is used to access the under-stimulated reservoir volume. Lantz et al. [47] reported the results from several refracturing treatments in the Bakken for 14 wells (cemented horizontal liner wells). During the refracturing operations, they noticed that the stress gradient dropped from 0.73 psi/ft. to 0.66 psi/ft. due to depletion (more details on stress changes can be found in [48]). The refracturing treatments were pumped at 50 bbl/min with 4 lb./gal sand concentration. They reported that the EUR estimate increased by more than 30%. The reader is referred to for more details about Refracturing candidate selection and treatment design [49].

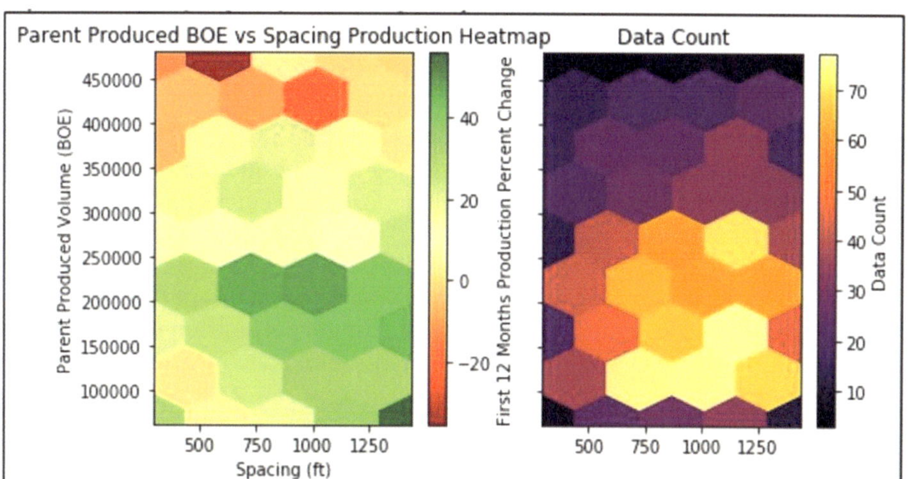

Figure 7.
Bakken wells spacing (primary and offset) and cumulative produced volume relationship to the first 12 month production percent change [44].

Dalkhaa et al. [50] reviewed 272 refracs that were conducted in the BPS. They used production rate changes, peak oil, change in GOR, and incremental EUR as metrics for the evaluation process. They found that open hole completions in the Bakken are under-stimulated compared to cased hole wells. They reported an incremental EUR of 340 MSTB in open hole refrac'ed. wells, and 175 MSTB in cased hole wells. They also noted that the decrease in GOR is an indicator of accessing new stimulated reservoir volume. The decrease was quantified for 90-days averaged daily production where open hole wells GOR decreased by 10% and cased hole wells did not decrease. They also conducted an economic analysis of refracs potential revenues and identified 400 potential wells as candidates for refracs. Following the study, Schmidt et al. [29] reported that 184 wells have been refrac'ed. between 2017 and 2021 with a decreasing trend from the highest levels in 2017.

Rostami et al. [51] conducted a detailed statistical analysis on the drilling direction of 7000 wells on well cumulative production and normalized production per lateral. The direction was then compared to the orientation of the least principal stress. They reported that wells drilled in the direction of the least principal stress were more economical compared to wells drilled in other directions. However, at most north or south leases, operators may find themselves with limited acreage, which results in drilling in the direction of maximum horizontal stress. These wells were more expensive to drill with higher cumulative production.

McClure et al. [52] reported results from a statistical 62 DFITs study to compare DFITs interpretation approaches. They report that a high percentage of Bakken DFITs does not show a clear closure which makes it challenging to estimate stress using the compliance method. This is related to the higher permeability of the play compared to other unconventional plays. The permeability ranges were sometimes outside the applicability of the compliance DFIT interpretation method. Dohmen et al. [53] reported results from 100 DFITs interpretations. The stress gradient was reported as 0.72–0.73 psi/ft. in the MB and 0.71–0.79 psi/ft. for TF. The pore pressure gradient was reported as 0.61–0.70 psi/ft. for MB and 0.72–0.73 for TF.

4.2 Pilot studies

Weddle et al. [54] evaluated the use of particulate diverters and increased number of perforation clusters per stage. In their attempt, they investigated the efficiency of their approach using radioactive tracers (RA). They investigated results from 27 wells that adapted the paper. The completion had 15 clusters per stage with 2 shots each at 0–180-degree phasing. The pumping rate was 80 bpm. The production results were compared based on 180 days of cumulative production. Their results reported that well production increased by 240% for high cluster efficiency (i.e., uniformity) compared to nearby wells. They also noted that the use of particulate diverters improved cluster efficiency. They note that the practitioner needs to account for settling velocity to determine the maximum perforation number. Their approach is limited to near wellbore uniformity (i.e. limited to measurable radioactive tracer distance).

Limited entry was first introduced by Murphy and Juch [55] and Lagrone and Rasmussen [56]. The method was mainly used to ensure equal flow rates between different perforations at different breakdown pressures. The approach is then adapted in unconventional plays to overcome the following issues [57]:

- Stress variability along the lateral length (90% of laterals are in the 750 range)

- Near-wellbore friction variability from cluster to another (P50 of 625 psi from step down tests)
- Stress shadow between cluster and from previous stages (formation elastic properties dependent)
- Fracture propagation variability in different clusters and elastic properties anisotropy
- Perforation friction changes due to erosion and perforation diameter range (approximately 500 psi)

Note that the values in brackets are values measured for the Bakken [57]. The pressure drop through perforation can be calculated as follows [58]:

$$\Delta P_p = \frac{0.2369 \times Q^2 \times \rho}{N_p^2 \times D_p^2 \times C_d^2} \quad (1)$$

Where ΔP_p is the perforation pressure (pressure drop across perforation (psi)). Q is the flow rate in bpm. ρ is the fracturing fluid density (lb/gal). N_p is the number of open perforations. D_p is the perforation diameter (in). C_d is the coefficient of discharge.

Weddle et al. [57] incorporated the use of the Extreme Limited Entry (XLE) in their designs to drive the number of clusters from 11 clusters per stage to 15 cluster per stage for the rate of 80 bpm. The pressure drop at perforations was designed at 2000 psi. They measured their success using data from fiber optic, radioactive tracers and production data. For a 9500 ft. lateral, they have drastically reduced the total number of stages from 50 to 27. The pilot project resulted in an average incremental production of 10%. The authors stress the importance of accounting for plug shifting that can cause leakage, thus, lower performance. The proppant transport and settling velocity estimations limit the maximum number of perforation clusters. The perforation orientation was mentioned, as some orientations provided better production compared to others. For more details on the perforation design, the reader can consult the following articles [59–62].

Lorwongngam et al. [63] investigated the design of XLE in the Williston Basin through a pilot field trial. They used pump pressure and step rate tests to ensure they attained high pressure. Radioactive tracer and fiber Distributed Acoustic Sensing/Distributed Temperature Sensing (DAS/DTS) and downhole camera were used to measure the uniformity of fluid distribution. They found out that the efficiency when using XLE design was independent of geological targets except for some cases. Using XLE design, the operator increased the cluster number to 15 per stage. The design was achieved using a limited entry pressure drop of 1500 psi. The design implements 2–3 perforations per cluster. In their field trial, they faced plug leakage issues that were identified using RA tracer data. Leakage problems were detrimental to the success of fracturing fluid uniform distribution.

Lorwongngam et al. [64] Improved the previous design by conducting field trials (XLE 2.0). They investigated the optimum shots per cluster, the minimum rate per cluster, and the maximum possible cluster number per stage. Their study was conducted for three years based on the Uniformity Index. A suite of fracture diagnostics was used to evaluate the eXtreme Limited Entry (XLE) design, including radioactive tracer, fiber optic, pressure step-down test, downhole camera, perforation acoustic imaging tools, lateral bottom hole gauges, in-well and offset fiber optic, production data. The results showed that using one shot per cluster can increase the cluster number to 20 per stage with 5 BPM for 4–1/2″ liner, and 6 BPM for a 5–1/2 liner. The pressure was 1900 psi. This reduces the total number of stages, thus, the rig time, leading to lower completion cost by 12%. **Table 3** reports the evolution of hydraulic fracturing design evolution.

Lorwongngam et al. [64] also noted that plug leakage and inefficient isolation can lead to lower cluster efficiency, thus, lower production. The authors also tried diverters in their trial; however, the XLE design was proven to give better production. They also suggest avoiding two perforation wireline runs.

Dohmen et al. [65] introduced the concept of Microseismic Depletion Delineation (MDD) for the first time. This technique was first introduced in the Bakken to delineate the depleted area from a producing parent well. The concept of the idea relies on the stress changes due to depletion. As the stress changes, the effective stress Mohr circle gets closer to the Coulomb failure envelope. Thus, fracture orientation in the critically stressed orientation will slip due to a change in pore pressure. Thus, the authors injected fluid in a depleted well and monitored the microseismic events resulting from the shear failure of critically stressed fractures to map the drained area. This concept was then used to plan infill drilling while accounting for depleted zones. The reader is referred to for more details on the approach [53, 66].

Cipolla et al. [67] reported the results of a field pilot of 6 laterals in which one is an observation lateral that was not stimulated drilled in the TF formation. The wells were equipped with SWPM and fiber optic to record FDI events and map far-field drainage. One MB well was equipped with fiber optic to measure the stage level uniformity and FDI occurrence from the stimulation of other wells. Another MB well was equipped with microseismic geophones to estimate the fracture geometry. Various sets of measurement and diagnostics techniques were used. The aim of these techniques is reported in **Table 4**.

Using their diagnostics, they reported fracture lengths ranging from 1900 ft. to 3800 ft. pumping the same volumes. They attributed this change to the effect of stress shadowing. They also analyzed FDI vs. fracture length data and Fracture length vs. injected volume to build empirical correlations and calibrate the fracture toughness. The uniformity index was used to calibrate the fluid distribution between clusters of the same stage. The pressure in the TF offset well was recorded during completion and production at several intervals of the lateral. The lateral pressure was used to

Design Type	Cluster per Stage (CPS)	Shot per Foot (SPF)	Estimated Cluster Spacing (ft)	Total Clusters/ well	Total Stages per Well
Baseline Original Design	3–6	3–6	35	280	36–42
XLE 1.0 Design of Experiment	6–15	2–3	33	300	30–32
XLE 2.0 Design of Experiment	8–20	1	30 >	300 <	25–27

Table 3.
Hydraulic fracturing design evolution [64].

	Comp. Efficiency	**Stim. Efficiency**	**Diversion**	**Frac-Geometry**	**Frac Morphology**	**Prod. Logging**	**Drainage Mapping**	**Well/FM Connectivity**
Fiber Optic								
Microseismic								
Pressure Gauges								
Oil/Water Tracers								
Proppant Tracers								
Camera								
Fingerprinting								
	Completion Design and Frac Characterization					Well Spacing and Frac Design		
	Development Optimization							

Table 4.
Diagnostics technics to record FDI events [67]. Red for high confidence measurement, and grey for lower confidence measurer.

assess the depletion of different fracture stage designs that intersect the observation wells at different pressure measurement locations. Some designs had better drainage than others. However, all designs of wells drilled in the MB drained from the TF formation at 450 ft. away from the primary well. Three fracturing designs were mainly evaluated to improve the production within one well. The third design had better connectivity to the reservoir for the longest period (note that the connectivity declined with time). The uniformity index from fiber data was highest at 14 clusters/stage and diminished with increasing cluster number. The uniformity index did not improve even with pod diverters (it got worse). This work resulted in a new concept called Augmented Drainage Development (ADD) [68]. ADD is an unstimulated well drilled in the middle of stimulated wells. This well drains from the induced fractures from offset wells to increase production. On the same project, Miranda et al. [69] reported an analysis of interference tests between wells in the MB and the TF formations using Chow Pressure Group Technique (CPG) [70]. They reported that the fracture system created in the DSU behaves as an interconnected system. The communication varies depending on the distance between wells and fracture design, this interference degrades with time. A MB well can drain up to 450 ft. from the lateral towards TF formation. These type of tests can be used to determine optimum well spacing and drainage performance.

4.3 Numerical modeling studies

Cipolla et al. [71] built a well-constrained model that was calibrated using MDD, microseismic, production logging tool, bottom hole gauge, RA tracers, and Image Logs geomechanics modeling and production data history matching to understand the impact of parent well depletion on the offset well fracture geometry. Their numerical study showed an effective fracture half-length of 900 ft., and a primary drainage of 400 ft. to both sides of the well. Wells drilled in the Bakken produced from TF and vice versa. A severe asymmetry in the offset well's fractures resulted from the depleted area near the primary well. In their approach, they used a fracture conductivity of 60md-ft for the propped fracture part and 4md-ft as a conductivity for the unpropped fracture part.

Lorwongngam et al. [72] modeled a pilot pad in the Bakken. To constrain their model, they used DFIT for stress and pore pressure estimation, microseismic for hydraulic fracturing geometry, MDD for drained area mapping, and RA tracer to quantify the uniformity index and the fracture propped half length of the fracture treatment. Oil and water tracers were used to quantify the connectivity between wells and bottom hole gauges and production data to history match the reservoir performance. They found that the effective fracture half-length is approximately 700–900 ft. for MB wells and 450 ft. for TF wells. The unpropped region contributes to the flow of hydrocarbons from the reservoir. Fracture conductivity ranges from 30md-ft in low stress to 9 md-ft for high stress propped fractures and 0.2 md-ft for unpropped fracture conductivity for high stress. The conductivity of the fracture in LBS is low, which limits the conductivity between wells in the TF and in the Bakken. This study's application was to increase well spacing and reduce the total number of wells per DSU.

Cipolla et al. [73] and Fowler et al. [74] conducted a numerical study to calibrate stress changes due to depletion (i.e., Biot coefficient). They used MDD to map the drained area, Microseismic to monitor fracture geometry, tiltmeter to measure fracture height, and time-lapse DFIT to measure stress and pore pressure changes due to depletion. They drilled two vertical wells next to a parent horizontal well that was depleted for 10 years. The DFITs were performed in the vertical wells to reduce the complexity of the problem. They reported that the connectivity distance between wells exceeds 1000 ft. The fracture from the vertical well grew asymmetrically towards the parent well. Used a Biot coefficient of 0.34 to match the MDD and FDI response. They compared two modeling approaches that resulted in similar results. First is the use of DFN modeling, and the second one using pre-existing planer fractures. The reader is referred to McClure et al. [75] for more details on planer fracture modeling.

Merzoug and Rasouli [76] built another model for the same previous case study using a hydraulic fracturing modeling approach to estimate fracture geometry. In their work, they found a different Biot coefficient of 0.7. They reported that the use of pre-existing fractures modeling approach changes the depletion zone and pressure distribution significantly. Their values align with results from laboratory work reported by Ling et al. [77] and Ma and Zoback [78]. Their study used proxy modeling to optimize the fracture geometry of offset wells. They reported that offset wells with wider spacing (1320 ft), tighter cluster spacing (26 ft), higher cluster numbers (20), and higher injection volumes (1900 bbl/cluster) resulted in better Net Present Values (NPV).

McClure et al. [79] reported results from a collaborative modeling study on the FDI effect in all major shale plays in North America. They analyzed three Bakken datasets as part of their study. The results showed that wells in the MB drain from the TF formation and vice versa. The FDI had a neutral effect on Bakken. The Bakken petroleum system is composed of a thin prolific zone. Thus, numerical modeling studies show that lateral fracture growth was important for better production in the Bakken, whereas vertical growth was not beneficial. The fracture spacing and well spacing were the dominating parameters in the production from the Bakken, with more production at tighter spacing. However, the decision depends on the interest of the operator in terms of optimizing net present value or discounted return on investment.

5. Artificial lift activity in the Bakken

According to Lane and Chokshi [80], the decline rates in shale plays, when production is not choked, can reach a staggering 60 to 80% within the first year. In the Bakken formation specifically, the average annual decline rate is as high as 69%. This significant drop is closely tied to the hydraulic fracturing process, which is essential for stimulating the low permeability formations characteristic of shale plays.

After the initial flowback of hydraulic fracturing fluid, a phase of high production rates typically follows as the fracture and surrounding rock start to drain. However, due to the low permeability, the migration of hydrocarbons to fractures is limited, leading to a swift drop in production rates [81, 82].

This rapid decline necessitates the implementation of artificial lift systems early in the well's lifespan, often just a few months after production begins. In some shale plays, artificial lift systems are even installed during the well commissioning phase, despite the expectation of natural flow during the initial months [81, 82].

Lolon et al. [41] analyzed the data for several wells completed in the Bakken/Three Forks formations and observed that artificial lift is required after only a few months of production due to the high decline rate. Ganpule et al. [83] arrived at a similar conclusion when they conducted a study examining the impact of completion practices on recoverable reserves in the Bakken/Three Forks formations. They found that the decision to switch to artificial lift systems was largely dependent on individual operator strategies, which were influenced by factors such as production targets and operational expenses. The study noted that some operators opted to transition to artificial lift systems relatively early, within 2 to 3 months of initial production, while others delayed this switch until 4 to 6 months into production. In some exceptional cases, high-performing wells did not require artificial lift systems until 18 to 24 months after the first production. This variability underscores the importance of operator-specific strategies in optimizing well performance and recoverable reserves.

5.1 Types of artificial lift systems

Based on a study by Eisner et al. [84], there are around 2 million operational oil wells globally, with over half - more than 1 million - utilizing some form of artificial lift. Sucker-rod pumps are the most common type of artificial lift, being used in over 750,000 wells. In the United States alone, approximately 350,000 wells employ sucker-rod pumps for lift.

A significant 80% of all U.S. oil wells are classified as stripper wells, producing less than 10 barrels per day, often with some water cut. Most of these stripper wells rely on sucker-rod pumps for lift. Looking at the non-stripper wells, which produce higher volumes, 27% use rod pumping, while 52% use gas lift. The remaining wells use other methods of lifting, including Electric Submersible Pumps (ESPs), hydraulic pumps, and other techniques.

In a more recent study, Pankaj et al. [85] provided a general breakdown of the artificial lift systems currently in use in U.S. onshore wells. They found that approximately 40% of wells starting with an artificial lift system use gas lift, while 36% employ ESPs. Rod lift is used in 13% of cases, plunger lift in 7%, and jet pumps in 4%.

In the Bakken formation, Patron et al. [86] stated that only ESPs and jet pumps (JP) can be implemented during the high-flow period. As production declines, a common practice is to move to sucker rod pumps (SRP), as a lower flow rate artificial lift method that allows a smooth transition with their ability to pump the required volumes from deep installations.

Britvar and Williams [87] reported that as of November 2016, Oasis Petroleum was operating on a substantial land area of 540,000 acres, with a primary focus on the Bakken and Three Forks formations. Most of the wells in operation were utilizing rod pumps for production. In addition, 30 wells were employing gas lift methods, while 35 wells were using ESPs for production. Britvar and Williams [87] also provided insights into the typical life cycle of artificial lift systems in the Williston Basin. Historically, operators have primarily used pumping units in this region. A typical well would employ a 912-pumping unit, which is moved to the location during the initial flowback, with rods being run after the well is loaded up. However, as completion designs evolved to move more fluid, some operators transitioned to long-stroke pumping units capable of significantly increasing fluid movement. The evolution of completion design in the basin necessitated the exploration of high-capacity lift systems to match improved well productivity. Artificial lift mechanisms such as ESPs and gas lift were identified as capable of moving considerably larger volumes at operating depths exceeding 10,000' TVD. The successful application of ESPs in Bakken and Three Forks wells has notably enhanced early life production rates, positively impacting total well economic returns.

In their study, Patron et al. [86] implemented an innovative artificial lift selection workflow to examine an unconventional well situated in the Three Forks formation. The process began with a pre-screening phase, where they evaluated the seven primary types of artificial lift systems based on both technical and economic criteria. Their analysis concluded that only Electric Submersible Pumps (ESP), Jet Pumps (JP), and Sucker Rod Pumps (SRP) were viable options for use throughout the well's lifespan. Interestingly, Gas Lift, a commonly used artificial lift system, was not included in their selection. This exclusion was primarily due to the substantial capital investment required to upgrade existing installations to support a gas lift injection infrastructure. The outcomes of their analysis are comprehensively presented in **Table 5**.

5.2 Field studies

In their study, Clark et al. [88] utilized production data from the North Dakota Industrial Commission to construct a database tracking production from all wells drilled and completed in the Bakken and Three Forks formations. They developed type curves based on rod pump artificial lift for two groups of wells, depending on the operating area.

Interestingly, they observed that wells with jet pumps tended to deviate positively from the type curve, indicating increased production. In fact, jet-pumped wells with the same completion strategy produced 11% more oil in the first year than rod-pumped wells.

The study further analyzed individual wells using hyperbolic decline to a terminal decline rate of 6%. The results showed that for the 10 wells with at least 12 months on a jet pump, the average increased production above the type curve for the first 12 months was 15.4 MBO. This resulted in an average incremental income of $1045 M per well, assuming 100% Working Interest, 78% Net Revenue Interest, $100/Barrel of Oil, and a differential of $13/Barrel of Oil.

Clark et al. [88] concluded that jet pumps can be an effective method for lifting Bakken and Three Forks wells in the Williston Basin, yielding superior production and economic results compared to rod-pumped wells. This is particularly true in higher water cut environments. They also noted that production results can be further optimized through automation controlling pump discharge to handle well slugging while optimizing drawdown.

Nickell and Treiberg [89] presented two case studies of Bakken wells that employ rod pumps, demonstrating the significant impact of optimizing a well's downtime on its operation. They utilized an autonomous control algorithm to modulate the idle time of the wells, aiming to reduce failures or increase production.

Artificial Lift Transitional	Artificial Lift Stage Late	Production Stage
Flowrate	700 B/D	200 B/D
Electrical Submersible Pump (ESP)	Applicable	Not Applicable
Jet Pump (JP)	Applicable	Not Applicable
Sucker Rod Pump (SRP)	Not Applicable	Applicable
Gas Lift (GL)	Not Applicable	Not Applicable
Progressive Cavity Pump (PCP)	Not Applicable	Not Applicable
Electrical Submersible Progressive Cavity Pump (ESPCP)	Not Applicable	Not Applicable
Plunger Lift (PL)	Not Applicable	Not Applicable

Table 5.
Results of artificial lift selection strategy for a well in the three forks formation [86].

In the first case study, a well experiencing fluid pound in the Bakken started with a downtime of 30 minutes, which the autonomous optimization algorithm increased to 100 minutes. This adjustment reduced the number of cycles per day from 30 to 8, without affecting production. This reduction in incomplete fillage pump strokes is expected to decrease failures on the downhole equipment, thereby increasing the well's runtime. The second case study also involved a Bakken well experiencing pump-off conditions with fluid pound. The well began with a downtime of 50 minutes, which the autonomous downtime optimization increased to 300 minutes. This increase led to a decrease in cycles per day from 25 to 5, again without reducing production or runtime. This adjustment is expected to reduce the number of fluid pound strokes by 36,500 per year, improving the overall health of the well and reducing failures.

Bestgen et al. [90] discussed the challenges faced by oil-producing wells in the Williston Basin (Bakken and Three Forks formations) that utilize Electrical Submersible Pump (ESP) lift. These wells encounter severe scale and corrosion issues due to high TDS brines and high shear and high-temperature operating conditions. The authors then presented the results of a field trial that compared the new combination product with a conventional scale/corrosion inhibitor. The new product provided a 73% increase in run time, and when used in conjunction with the pre-conditioning corrosion inhibitor treatment, the run time showed a 100% increase. The paper concludes that any extension in ESP run time gained through mechanical or chemical means offers improved well production, lower operating expenses, and improved profitability to the producer.

Orji et al. [91] conducted a study on the optimization of unconventional wells operated by Hess in the Bakken that are using sucker rod lift. Their research revealed that the optimization process necessitates a delicate balance between reservoir inflow, gas separation (pump efficiency), and structural integrity parameters. They utilized a dimensionless pressure and dimensionless time (pDtD) approach, which proved effective in capturing complex physical phenomena associated with unconventional wells, such as transient, multi-stage fractures, and superposition effects. This approach was found to be simpler and faster to calibrate compared to traditional numerical simulator history matching exercises, making it particularly useful for scenarios involving a high number of wells.

The authors clarified that their approach is intended to complement, not replace, classical reservoir engineering work by addressing short-term operational challenges. They also highlighted the time-dependent nature of unconventional wells, indicating that the modeled production potential is a moving target, and thus, optimization recommendations should be implemented as swiftly and safely as possible. The study also underscored the importance of ensuring the structural integrity and gearbox loading of the pumping unit, the adequacy of communication and control systems on the pumping unit, and the accuracy of the metering and allocation method on a given site for successful optimization.

Rankin et al. [92] performed a case study analysis to evaluate improvements and profitability achieved with superior completions in the Bakken/Three Forks horizontal wells. A side observation from this study is that wells completed with multiple Plug and Perf stages flowed naturally for longer periods compared to most wells completed with fewer frac sleeves (3 out of 4 wells), which required the installation of pumping units earlier in their life cycle.

Wells F, G, and H, which utilized Plug & Perf methods with ceramic proppant, show higher initial production rates but also experience a steep decline. This rapid decline necessitates the earlier installation of pumping units as indicated by the proximity of the stars to the y-axis. In contrast, Wells I, J, K, and L employed Frac Sleeves with white sand proppant, demonstrating lower initial production rates but a more gradual decline. Notably, the cumulative production for these wells remains competitive with that of the higher-performing wells over time, illustrating the trade-off between initial production vigor and long-term decline rates.

According to a study by Yuan et al. [93], the impact of well trajectory on the production performance of ESP-lifted shale wells was investigated. The study utilized a representative set of field data to model the shale well with an ESP installed. The wells under study had similar reservoir characteristics but varied in their well trajectories, including smooth and highly sinuous toe-up and toe-down trajectories. A transient dynamic multiphase simulator was used to integrate ESP performance and wellbore models over the life of a shale well. The study found that well trajectory does not significantly impact the production performance of an ESP-lifted well, even under conditions of slugging flow. The research also highlighted the importance of using a transient wellbore simulator to capture the continuous, long-period transient process inherent in the production of unconventional shale wells. The study concluded the workflow described could be used to evaluate selected ESP performance and timing for installation and withdrawal for a particular well over its whole life cycle.

6. Application of machine learning in artificial lift

Pennel and Hsiung [94] presented a workflow to diagnose artificial lift problems using machine learning. They trained their models using production data for over 1 year from operating wells in the Bakken. Their objective was to predict the operational problems of sucker rod pumps and gas lift systems including tubing and pump failure along with the suboptimal performance of sucker rod pumps and gas injection. The data consisted of time series sensor and controller data for 800 wells, reviewed by experts in each lift type to identify periods of failure and sub-optimal performance.

Freeman et al. [95] embarked on a study to optimize unconventional wells in the Bakken using an Internet of Things (IoT) device with high-performance computational capabilities. The researchers developed an IoT device capable of real-time analysis and higher-order mathematics, which was connected to a cloud-based analytics software platform. This technology was deployed on 50 representative wells in the Bakken, with the device connected to the legacy rod pump controller via Modbus connection. The results from the IoT device showed immediate differences in key downhole parameters when compared to the traditional rod pump controller.

The study found that the higher-accuracy physics-based inputs fed into machine learning algorithms, which dynamically classified wells into key operating states of under-pumping, over-pumping, and dialed-in. The results showed that for wells that were under-pumping, Equinor was able to increase oil production by up to 33%. For wells that were over-pumping, Equinor was able to decrease the number of strokes by 11% and increase pump efficiency by 14%. The study concluded that the vision of autonomous well operations is possible to implement, and the investment in modern optimization technology provides lasting, repeatable value through many operational parameters.

Table 6, as derived from the study by Freeman et al. [95], presents a concise summary of the production efficiency gains achieved through the implementation of an optimization strategy on unconventional wells in the Bakken formation. The metrics depicted in the table provide a quantitative comparison of well performance before and after the application of the optimization techniques.

Metric	Average
Total production before (bpd)	1842
Total production after (bpd)	2447
Production uplift	605
Production increase on under-pumping wells (%)	32.8%

Table 6.
Summary statistics showing efficiency gains without impacting production (after Freeman et al. [95]).

Specifically, the table shows that the average total oil production before optimization was 1842 barrels per day (bpd). After optimization, the production increased to an average of 2447 bpd, indicating a significant uplift of 605 bpd. This demonstrates a clear enhancement in well productivity attributable to the optimization process.

Moreover, a notable metric of improvement is the average production increase of 32.8% on wells that were initially under-pumping. This statistic underscores the potential for production gains by addressing inefficiencies in the pumping process and optimizing operational parameters. The table effectively highlights the tangible benefits of applying targeted optimization techniques in terms of increased oil production and improved well performance, reinforcing the value of such interventions in the context of IOR strategies.

7. Discussion

The drilling operations within the Williston Basin have yielded valuable insights that can guide the industry towards more effective and efficient extraction methods. The implementation of advanced drilling technologies, such as high-performance motors and real-time downhole drilling data, has significantly enhanced drilling efficiency and oil production. Moreover, the accuracy of geological placement through precise geosteering techniques has proven instrumental in minimizing errors and maximizing production.

Horizontal well completions, with their higher production rates compared to traditional vertical completions, have clearly established themselves as the way forward. When paired with effective drilling practices such as using cemented liner wells with diversion, longer laterals, and cleaner fluids, substantial production improvements have been achieved.

Attention to the vibration issue is critical, and further research into vibration measurement and mitigation is a pressing need for enhancing operational productivity. Likewise, minimizing sliding during operations through motor-driven rotary steerable systems enhances hole quality and offers better directional control.

The role of infrastructure, especially in terms of accommodating produced gas and oil, emerged as a central factor affecting drilling operations. Similarly, the availability of skilled labor directly impacts the productivity of drilling sites. As high-yielding core areas near exhaustion, identifying potential new drilling locations has become a crucial step in maintaining production levels.

Adapting to innovative practices like the use of wet shoetrack completions can usher in cost savings and operational optimization. Furthermore, the adoption of advanced well completion techniques, such as expandable liner hangers and MPD, are critical in overcoming the challenges associated with long lateral drilling and completion of horizontal wells.

Given that drilling operations are energy-intensive, strategies aimed at improving energy efficiency are necessary to mitigate the environmental impact. Lastly, the ability to cope with industry fluctuations due to economic or other crises is integral to maintaining operational continuity and productivity.

The evolution of hydraulic fracturing design was the result of a combination of science-oriented trial-and-error projects. Operators started with small hydraulic fracturing designs using cross-linked gel. The trend then changed to increasing fluid volumes and mass. The optimization went from a well-basis improvement into a DSU optimization where the recoveries are quantified on a well basis. The increased number of infill wells caused concern for operators because of overlapping drainage areas. However, In Bakken, a trend was noted. Old wells treated with lower volumes showed an uplift in production, whereas the newer wells were not influenced by FDI. Some operators attempted to reduce the number of stages. This resulted in a decrease in costs while keeping production at higher levels. This reduction in the number of stages is associated with an increase in the number of perforation clusters. This was made possible using high limited entry pressures and High Viscosity Friction Reducer fluids with a minimum rate of 5–6 bpm per cluster. Other operators keep the high number of stages with tighter well spacing. The two approaches are reported in Appendix A for two different operators. The field testing and numerical modeling approaches showed that the wells drilled in the MB drain from the TF formations and vice versa. The wells drilled in the direction of the least principal stress were more optimum in recovery because they create the largest surface area. Refrac operations were executed in several wells in the Bakken. However, the success rate was limited. The fracture geometry was estimated to propagate upward towards the Lodgepole and downward towards the lower TF. The fractures drain 450 ft. (measured) and a total drainage of 700–900 ft. (modeled) away from the lateral. The estimated stress gradient is around 0.7 psi/ft. with a Biot coefficient of about 0.34–0.7.

The rapid decline in oil production in shale plays like the Bakken formation requires the implementation of artificial lift systems early in the well's lifespan. Often, artificial lift systems are installed just months after production begins, despite the expectation of natural flow. The timing of this switch can vary greatly depending on individual operator strategies, influenced by production targets and operational expenses. Regarding the types of artificial lift systems used, global studies have shown that sucker-rod pumps are the most common type, followed by gas lift in high volume wells. Electric Submersible Pumps (ESPs) and jet pumps are typically implemented during the high-flow period in the Bakken formation. As production declines, a common practice is to transition to sucker rod pumps.

Artificial lift systems and their implementation strategies have significantly evolved over time to keep up with changing well productivity. For instance, high-capacity lift systems such as ESPs and gas lift have been identified as capable of moving considerably larger volumes, and have significantly enhanced early life production rates, thus positively impacting total well economic returns. The optimization of unconventional wells necessitates a delicate balance between reservoir inflow, gas separation, and structural integrity parameters. Techniques like dimensionless pressure and dimensionless time approach can capture complex physical phenomena associated with unconventional wells and offer simple, swift solutions to address short-term operational challenges.

Field studies on the Bakken and Three Forks formations have shown interesting results. Jet-pumped wells tend to deviate positively from the type curve, indicating increased ultimate production. Autonomously controlled rod-pump wells can significantly reduce failures or increase production by optimizing downtime. Furthermore, ESP-lifted wells can greatly increase their runtime by combating severe scale and corrosion issues through innovative chemical treatments.

Machine learning has been increasingly applied to diagnose artificial lift problems and optimize unconventional wells. Machine learning models trained on production data can predict operational issues like tubing and pump failures, improving the overall health of the well and reducing failures. In parallel, the advent of Internet of Things (IoT) devices in the oil and gas industry allows real-time analysis and higher-order mathematics, which can classify wells into key operating states of under-pumping, over-pumping, and dialed-in, and optimize their operational parameters accordingly.

8. Conclusion

This chapter summarizes the best practices performed by operators in the Bakken Petroleum System. It reports the recent findings and operations for completing wells in the unconventional play. It can serve as a roadmap for people starting to work in the Bakken or a lesson learned paper for people around the world to continue from current best practices in the Bakken at the time the paper was published. The Drilling operations had accelerated dramatically bringing the cost to very low levels. The hydraulic fracturing design was changes several time to improve well productivity and different concepts about fracture geometry have been changed through time. The artificial lift application approaches have been tested in the Bakken and had different performance under different conditions. The optimization process changed from a well basis into a DSU basis. The Bakken evolution was made possible thanks to field trial, since projects supported with numerical simulation and statistical studies. Learning from previous practices is what drove the completion evolution and it will continue in the future.

Nomenclature

ADD	Augmented Drainage Development
AL	Artificial Lift
BPS	Bakken Petroleum System
BS	Ball and Sleeve
Cmt	Cemented
Comp.	Completion
CPG	Chow Pressure Group Technique
DAS/DTS	Distributed Acoustic Sensing/Distributed Temperature Sensing
DFIT	Diagnostic Fracture Injection Test
DSU	Drilling and Spacing Unit
ESP	Electrical Submersible Pump
ESPCP	Electrical Submersible Progressive Cavity Pump
FDI	Fracture Driven Interaction
Gel	XL Gel
GL	Gas Lift
GOR	Gas Oil Ratio
HFE	High-Frequency Extender

Hyb.	Hybrid Slickwater/Gel
IOR	Improved Oil Recovery
IoT	Internet of Things
JP	Jet Pump
MD	Measured Depth
MDD	Microseismic Depletion Delineation
MWD/LWD	Measurement While Drilling/Logging While Drilling
NER	Net Energy Return
NPV	Net Present Value
OH	Open Hole
PCP	Progressive Cavity Pump
PL	Plunger Lift
PNP	Plug and Perf
Prop.	Proppant volume/ length
RA	Radioactive Tracer Analysis
ROP	Rate of Penetration
SP	Swell Packer
SRP	Sucker Rod Pump
SS	Sliding Sleeve
SW	Slick Water
TF	Three Forks
TVD	True Vertical Depth
Vol.	Volumes
WC	Water Cut
XLE	Extreme Limited Entry

Appendix

	Well #	Frac-Vol. (bbls)	Sand Lb (lb)	#Stages	Max Inj Rate (bpm)	Max Treatment Pressure	lat length	Well Spacing (ft)	Cum Oil	Cum Water	Cum Gas	IP Oil	IP Water	IP G as	Date	Bench
Operator M	37,778	152,832	8,492,476	40	108.8	9551	9968	700	326,894	192,890	564,961	1833	1006	2346	7/7/2021	MB
	37,777	159,275	8,507,435	40	110.8	9358	9890	450	367,194	311,225	690,061	2188	1757	2591	7/7/2021	TF Bench 1
	37,774	153,367	8,499,679	40	109.7	9883	9663	450	333,552	165,385	587,024	2391	1048	2944	6/15/2021	MB
	37,773	157,379	8,502,985	40	111	9548	9635	450	384,685	288,834	641,572	2488	1910	3034	6/15/2021	TF Bench 1
	37,772	155,647	8,489,220	40	111.5	9868	9940	450	270,302	310,827	601,335	1145	1486	1421	6/15/2021	TF Bench 1
	37,787	156,965	8,477,568	40	106	9595	9544	450	338,912	329,079	597,432	1351	1657	1504	8/6/2021	TF Bench 1
	37,786	147,605	8,332,270	40	69.3	9554	9625	450	326,935	188,470	556,636	1498	880	1811	9/2/2021	MB
	37,740	200,161	12,479,510	46	92.4	9474	11,290	500	406,094	332,738	318,517	1615	1488	1089	8/25/2021	MB
Operator H	37,979	261,486	16,003,347	20	123	8889	10,918	660	299,084	372,992	475,557	1289	1833	1531	5/7/2021	MB
	37,978	261,212	15,993,640	20	124	9250	10,903	660	327,738	383,660	578,005	1354	1970	2368	4/28/2021	MB
	37,128	185,939	9,733,347	26	79	9074	9896	660	142,013	220,781	287,803	833	1518	1455	7/3/2021	MB
	37,129	171,741	10,030,421	24	91	9261	9921	660	160,466	239,811	344,453	908	1564	1204	7/10/2021	MB
	37,130	171,568	10,027,942	24	91	8902	9912	660	172,154	208,757	385,018	835	689	866	7/7/2021	MB
	37,706	223,681	10,018,586	24	93	9410	9962	450	186,170	342,117	460,124	689	1852	1343	9/27/2021	TF Bench 1
	38,310	316,661	15,034,262	26	90.9	8851	9893	1100	233,524	278,600	482,183	1079	865	1616	11/7/2021	MB
	36,520	203,602	9,964,933	24	91	9527	9928	450	226,118	253,462	447,761	891	1224	1471	10/8/2021	TF Bench 1

Author details

Ahmed Merzoug[1*], Aimen Laalam[2], Lynn Helms[3], Habib Ouadi[4], John Harju[4] and Olusegun Stanley Tomomewo[5]

1 Petroleum and Geosystems Engineering Department, University of Texas at Austin, USA

2 Petroleum Engineering Department, University of North Dakota, USA

3 North Dakota Department of Mineral Resources, USA

4 Energy and Environmental Research Center (EERC), University of North Dakota, USA

5 University of North Dakota, USA

*Address all correspondence to: amerzoug@utexas.edu

References

[1] Berman A. The Decline And Fall Of North Dakota's Bakken Oil Play. Forbes. 2020. Available from: https://www.forbes.com

[2] U.S. Energy Information Administration. Drilling Productivity Report (DPR). 2023. Available from: https://www.eia.gov/petroleum/drilling/

[3] Sorensen JA, Schmidt DD, Smith SA, Bailey TP, Mibeck BAF, Harju JA. Factors Affecting Successful Oil Production in the Bakken Formation, North Dakota. 2010. Available from: https://undeerc.org/bakken/pdfs/ JAS_Bakken_May10.pdf

[4] Drilling Edge. North Dakota Permits, Production, Wells & Operators. 2023. Available from: https://www.drillingedge.com/north-dakota

[5] North Dakota Industrial Commission. North Dakota general statistics. 2023. Available from: https://www.dmr.nd.gov/oilgas/stats/statisticsvw.asp

[6] Miller G, Lindsay G, Baihly J, Xu T. Parent well refracturing: Economic safety nets in an uneconomic market. In: Society of Petroleum Engineers - SPE Low Perm Symposium. Denver, Colorado, USA. May 2016. DOI: 10.2118/180200-ms

[7] Acuña JA. Straightforward representative fluid flow models for complex fracture networks in unconventional reservoirs. In: Paper presented at the SPE/AAPG/SEG Unconventional Resources Technology Conference. Houston, Texas, USA. July 2018. DOI: 10.15530/urtec-2018-2876208

[8] Latrach A, Merzoug A, Abdelhamid C, Mellal I, Rabiei M. Identification and quantification of the effect of fracture-driven interactions on production from parent and child Wells in Williston Basin. In: Unconventional Rescources Technology Conference. Denver, Colorado, USA: SPE/AAPG/SEG Unconventional Resources Technology Conference; June 2023. pp. 1-23. DOI: 10.15530/urtec-2023-3868861

[9] LeFever JA. History of oil production from the Bakken Formation, North Dakota. In: Hansen WB, editor. Williston Basin Symposium. Guidebook to Geology and Horizontal Drilling of the Bakken Formation: Montana Geological Society. Billings, Montana. 1991. pp. 3-18

[10] Pitman JK, Price LC, LeFever JA. Diagenesis and fracture development in the Bakken Formation, Williston Basin: Implications for reservoir quality in the middle member. In: U.S. Geological Survey Professional Paper. North Dakota, USA. 2001. DOI: 10.3133/pp1653

[11] Clark A. The Bakken Boom: A Modern-Day Gold Rush. The Oil Drum; 2011. Available from: http://theoildrum.com/node/8697 [Accessed: August 10, 2023]

[12] Helms L. North Dakota Department of Mineral Resources Monthly Report. North Dakota, USA. 2012

[13] Gaswirth SB, Marra KR, Cook TA, Charpentier RR, Gautier DL, Higley DK, et al. Assessment of Undiscovered Oil Resources in the Bakken and Three Forks Formations, Williston Basin Province, Montana, North Dakota, and South Dakota, 2013. U.S. Geological Survey Fact Sheet 2013-3013; 2013. p. 4. Available from: https://pubs.usgs.gov/fs/2013/3013

[14] Djurisic A, Binnion A, Taglieri A, Thompson J, Menge M, Fleischhacker C,

et al. Williston Basin - A history of continuous performance improvements drilling through the Bakken. In: IADC/SPE Drilling Conference and Exhibition. New Orleans, Louisiana, USA. February 2010. DOI: 10.2118/128720-MS

[15] Han G, Davila WA, Magnuson EC, Azizov AA. Practical directional drilling techniques and MWD technology in Bakken and upper three forks formation in Williston Basin North Dakota to improve efficiency of drilling and well productivity. In: SPE Middle East Unconventional Gas Conference and Exhibition (Paper No. SPE 163957). Muscat, Oman: Baker Hughes; 2013

[16] Pearson CM, Griffin LG, Strickland SL, Weddle PM. Twelve years and twelve thousand multi-stage horizontal wells in the Bakken - how is industry continuing to increase the cumulative production per well? In: SPE International Hydraulic Fracturing Technology Conference and Exhibition. Muscat, Oman. 2018. DOI: 10.2118/191455-18IHFT-MS

[17] Lolon EP, Cipolla CL, Weijers L, Hesketh RE, Grigg MW. Evaluating horizontal well placement and hydraulic fracture spacing/conductivity in the Bakken formation, North Dakota. In: SPE Annual Technical Conference and Exhibition. New Orleans, Louisiana. October 2009. DOI: 10.2118/124905-MS

[18] Johnson CC, Courrege DW. Advances in openhole packer/sleeve completions coupled with multilaterals provide cost-effective solutions. In: SPE Annual Technical Conference and Exhibition (Paper No. SPE-135268-MS). Florence, Italy. 2010. DOI: 10.2118/135268-MS

[19] Southcott A, Harper H. 3-D seismic proves its value in Bakken geosteering. In: SPE/AAPG/SEG Unconventional Resources Technology Conference (Paper No. URTEC-1922656-MS). Denver, Colorado, USA. 2014. DOI: 10.15530/URTEC-2014-1922656

[20] Brandt AR, Yeskoo T, Vafi K. Net energy analysis of Bakken crude oil production using a well-level engineering-based model. Energy. 2015;**93**(Part 2):2191-2198. DOI: 10.1016/j.energy.2015.10.113

[21] Gutierrez CS, Parker T, Hinz D, Shearer E, Carson D. Wellbore placement in the middle Bakken formation using an azimuthal gamma-Ray measurement. In: SPE/AAPG/SEG Unconventional Resources Technology Conference (Paper No. URTEC-2461299-MS). San Antonio, Texas, USA. 2016. DOI: 10.15530/URTEC-2016-2461299

[22] Lim JC, Cortes P, Smith L. An integration of drilling/completion technologies for maximizing long-term productivity. In: Paper Presented at the SPE/IADC Middle East Drilling Technology Conference and Exhibition; Abu Dhabi, UAE. 2016. DOI: 10.2118/178237-MS

[23] Veazey MV. Whiting Boosts Williston Basin Drilling Inventory. Rigzone. 2021. Available from: https://www.rigzone.com/news/whiting_boosts_williston_basin_drilling_inventory-22-jul-2021-166008-article/

[24] Halliburton. Williston Basin Operator Increases Drilling Efficiency Via Intelligent Bit Technology. 2019. Available from: https://cdn.brandfolder.io/3I2CY2XL/as/mhgmh84b8kv8krmmbrb3fgr/williston-basin-operator-increases-drilling-efficieny.pdf

[25] Jerrard M, Osborne P, Stark P. Fiber optics improve cement performance and drilling cost in the Bakken. In: IADC/SPE International Drilling Conference and Exhibition. Galveston, Texas, USA. 2022. DOI: 10.2118/208673-MS

[26] Halliburton. Operator Uses Polymer Sealant for Improved Wellbore Stability in the Bakken Shale. 2022a. Available from: https://cdn.brandfolder.io/3RYPUX6K/at/r9pp7cvtsh9ncgjw3h66jhth/Operator_Uses_Gummy_Polymer_Sealant_for_Improved_Wellbore_Stability_in_the_Bakken_Shale.pdf

[27] Halliburton. Faster Well Delivery With Enhanced High-Torque, High Flow Motors. 2022b. Available from: https://brandfolder.io/item/apply/copy/h014336

[28] Gundersen A, Gomez R, Kadler C, Broussard T, Gandomkar A. Intentional wet shoetrack completions for unconventional oil and gas resources: Methods and techniques. In: SPE Oklahoma City Oil and Gas Symposium. Oklahoma City, Oklahoma, USA. April 2023. DOI: 10.2118/213081-MS

[29] Schmidt D, Min KS, Guillot S. Parent Well Strategies in the Bakken, Trends and Outcomes. SPE Hydraulic Fracturing Technology Conference and Exhibition. Texas, USA: The Woodlands; January 2023. DOI: 10.2118/212373-MS

[30] Ouadi H, Laalam A, Hassan A, Chemmakh A, Rasouli V, Mahmoud M. Design and performance analysis of dry gas fishbone wells for lower carbon footprint. Fuels. 2023b;**4**(1):92-110. DOI: 10.3390/fuels4010007

[31] Ouadi H, Mishani S, Rasouli V. Applications of underbalanced fishbone drilling for improved recovery and reduced carbon footprint in unconventional plays. Petroleum & Petrochemical Engineering Journal (PPEJ). 2023a. DOI: 10.23880/ppej-16000331

[32] Reddy S, Pitcher J. Application of 3-D Geosteering capabilities in geologically-complex shale. In: SPE Americas Unconventional Resources Conference (Paper No. SPE-153160-MS). Pittsburgh, Pennsylvania, USA. 2012. DOI: 10.2118/153160-MS

[33] Bassarath WD, Maranuk CA. Application of targeted bit speed (TBS) technology to optimize Bakken shale drilling. In: SPE/IADC Drilling Conference and Exhibition (Paper No. SPE/IADC 163406). Amsterdam, The Netherlands: Weatherford; 2013

[34] McCormick J, Brown C, Walters G. High compression expandable liner hanger case history in the Bakken shale: Pre-job planning, real time monitoring and post job analysis. In: IADC/SPE Asia Pacific Drilling Technology Conference (Paper No. SPE-170469-MS). Bangkok, Thailand. 2014. DOI: 10.2118/170469-MS

[35] Parayno G, Peacock S, Connolly B. Revitalizing the bakken with managed pressure drilling. In: SPE/IADC Drilling International Drilling Conference and Exhibition. The Hague, The Netherlands. March 2019. Available from: https://www.onepetro.org/conference-paper/SPE-194114-MS

[36] Chinander H. Bakken Oil Production Lagging Despite Higher Prices. Federal Reserve Bank of Minneapolis. 2021. Available from: https://www.minneapolisfed.org/article/2021/bakken-oil-production-lagging-despite-higher-prices

[37] Tobben S. North Dakota's Bakken Shale Field is Holding Back Oil Output. Bloomberg. 2023. Available from: https://www.bloomberg.com/news/articles/2023-02-07/aging-shale-field-in-north-dakota-is-holding-back-us-oil-output?leadSource=uverify%20wall

[38] Western Dakota Energy Association. Helms Wants More Three-Mile Laterals.

2022. Available from: https://www.ndenergy.org/Newsletter/Fall-Colors/Helms-Wants-More-Three-Mile-Laterals

[39] Jean R. Labor, Lack of Infrastructure Are Taking the Top off North Dakota's Oil and Gas Recovery. Williston Herald. 2023. Available from: https://www.willistonherald.com/news/oil_and_energy/labor-lack-of-infrastructure-are-taking-the-top-off-north-dakotas-oil-and-gas-recovery/article_68672a6c-935e-11ec-a69e-df734464fe8d.html

[40] Griffin L, Pearson CM, Strickland S, McChesney J, Wright C, Mayer J, et al. The Value Proposition for Applying Advanced Completion and Stimulation Designs to the Bakken Central Basin. SPE Annual Technical Conference and Exhibition. New Orleans, Louisiana, USA. September 2013. DOI: 10.2118/166479-MS

[41] Lolon E, Hamidieh K, Weijers L, Mayerhofer M, Melcher H, Oduba O. Evaluating the relationship between well parameters and production using multivariate statistical models: A middle Bakken and three forks case history. In: Paper presented at the SPE Hydraulic Fracturing Technology Conference, The Woodlands, Texas, USA, February 2016. 2016a. DOI: 10.2118/179171-MS

[42] Pearson CM, Fowler G, Stribling KM, McChesney J, McClure M, McGuigan T, et al. Near-Wellbore Deposition of High Conductivity Proppant to Improve Effective Fracture Conductivity and Productivity of Horizontal Well Stimulations. SPE Annual Technical Conference and Exhibition, Virtual. October 2020. DOI: 10.2118/201641-MS

[43] Taghavinejad A, Brown C, Ostadhassan M, Liu B, Hadavimoghaddam F, Sharifi M. A real-world impact of offset frac-hits by rate transient analysis in the Bakken and three forks, North Dakota, USA. Journal of Petroleum Science and Engineering. 2022;**208**(PE):109710. DOI: 10.1016/j.petrol.2021.109710

[44] Cozby J, Sharma M. Parent-Child Well Relationships Across US Unconventional Basins: Learnings from a Data Analytics Study. SPE Hydraulic Fracturing Technology Conference and Exhibition, The Woodlands, Texas, USA. February 2022. DOI: 10.2118/209171-MS

[45] Xu T, Lindsay G, Zheng W, Yan Q, Patron KE, Alimahomed F, et al. Advanced modeling of production induced pressure depletion and well spacing impact on infill wells in spraberry, Permian basin. In: Proceedings - SPE Annual Technical Conference and Exhibition. Dallas, Texas, USA. September 2018. DOI: 10.2118/191696-ms

[46] Gupta I, Rai C, Devegowda D, Sondergeld CH. Fracture hits in unconventional reservoirs: A critical review. SPE Journal. 2021;**26**(1):412-434. DOI: 10.2118/203839-PA

[47] Lantz T, Greene D, Eberhard M, Pershall R. Refracture Treatments Proving Successful in Horizontal Bakken Wells: Richland Count, Montana. Rocky Mountain Oil & Gas Technology Symposium. Denver, Colorado, USA. April 2007. DOI: 10.2118/108117-MS

[48] Agrawal S, Sharma MM. Impact of pore pressure depletion on stress reorientation and its implications on the growth of child well fractures. In: SPE/AAPG/SEG Unconventional Resources Technology Conference. Houston, Texas, USA. July 2018. DOI: 10.15530/urtec-2018-2875375

[49] Kong L, Ostadhassan M, Tamimi N, Samani S, Li C. Refracturing: Well selection, treatment design, and lessons learned—A review. Arabian

Journal of Geosciences. 2019;**12**(4):117. DOI: 10.1007/s12517-019-4281-8

[50] Dalkhaa C, Azzolina NA, Chakhmakhchev A, Kurz BA, Sorensen JA, Gorecki CD, et al. Refracturing in the Bakken - An analysis of data from across North Dakota. In: SPE/AAPG/SEG Unconventional Resources Technology Conference. Houston, Texas, USA. June 2022. DOI: 10.15530/urtec-2022-3723732

[51] Rostami E, Boness N, Zoback MD. Significance of well orientation on cumulative production from Wells in the Bakken region. In: SPE/AAPG/SEG Unconventional Resources Technology Conference Virtual. July 2020. DOI: 10.15530/urtec-2020-2813

[52] Mcclure M, Fowler G, Picone M. Best Practices in DFIT Interpretation: Comparative Analysis of 62 DFITs from Nine Different Shale Plays. In: SPE International Hydraulic Fracturing Technology Conference & Exhibition. Muscat, Oman. January 2022. DOI: 10.2118/205297-MS

[53] Dohmen T, Zhang J, Barker L, Blangy JP. Microseismic magnitudes and b -values for delineating hydraulic fracturing and depletion. SPE Journal. 2017a;**22**(05):1624-1634. DOI: 10.2118/186096-PA

[54] Weddle P, Griffin L, Pearson CM. Mining the Bakken: Driving Cluster Efficiency Higher Using Particulate Diverters. In: SPE Hydraulic Fracturing Technology Conference and Exhibition, The Woodlands, Texas, USA, January 2017. DOI: 10.2118/184828-MS

[55] Murphy WB, Juch AH. Pin-point Sandfracturing - a method of simultaneous injection into selected sands. Journal of Petroleum Technology. 1960;**12**(11):21-24. DOI: 10.2118/1415-G

[56] Lagrone KW, Rasmussen JW. A new development in completion methods-the limited entry technique. Journal of Petroleum Technology. 1963;**15**(07):695-702. DOI: 10.2118/530-PA

[57] Weddle P, Griffin L, Pearson CM. Mining the Bakken II – Pushing the Envelope with Extreme Limited Entry Perforating. In: SPE Hydraulic Fracturing Technology Conference and Exhibition, The Woodlands, Texas, USA, January 2018. DOI: 10.2118/189880-MS

[58] Cramer DD. The Application of Limited-Entry Techniques in Massive Hydraulic Fracturing Treatments. In: SPE Production Operations Symposium, Oklahoma City, Oklahoma. March 1987. DOI: 10.2118/16189-MS

[59] Dontsov E. A Model for Proppant Dynamics in a Perforated Wellbore. International Journal of Multiphase Flow. 2023;**167**:104552. Available from: http://arxiv.org/abs/2301.10855

[60] Liu X, Wang J, Singh A, Rijken M, Wehunt D, Chrusch L, et al. Achieving near-uniform fluid and proppant placement in multistage fractured horizontal Wells: A computational fluid dynamics modeling approach. SPE Production & Operations. 2021;**36**(04):926-945. DOI: 10.2118/204182-PA

[61] Snider P, Baumgartner S, Mayerhofer M, Woltz M. Execution and Learnings from the First Two Surface Tests Replicating Unconventional Fracturing and Proppant Transport. In: SPE Hydraulic Fracturing Technology Conference and Exhibition. The Woodlands, Texas, USA. February 2022. DOI: 10.2118/209141-MS

[62] Wu C-H, Sharma MM. Effect of Perforation Geometry and Orientation on Proppant Placement in Perforation

Clusters in a Horizontal Well. SPE Hydraulic Fracturing Technology Conference. The Woodlands, Texas, USA. February 2016. DOI: 10.2118/179117-MS

[63] Lorwongngam (Ohm) A, Wright S, Hari S, Butler E, McKimmy M, Wolters J, et al. Using multidisciplinary data gathering to evaluate eXtreme limited entry completion design and improve perforation cluster efficiency. In: SPE/AAPG/SEG Unconventional Resources Technology Conference, Virtual. July 2020. DOI: 10.15530/urtec-2020-2796

[64] Lorwongngam AO, McKimmy M, Oughton E, Cipolla C. One Shot Wonder XLE Design: A Continuous Improvement Case Study of Developing XLE Design in the Bakken. In: SPE Hydraulic Fracturing Technology Conference and Exhibition. The Woodlands, Texas, USA. January 2023. DOI: 10.2118/212358-MS

[65] Dohmen T, Zhang J, Li C, Blangy JP, Simon KM, Valleau DN, et al. A new surveillance method for delineation of depletion using microseismic and its application to development of unconventional reservoirs. In: SPE Annual Technical Conference and Exhibition. Vol. 3. New Orleans, Louisiana, USA. September 2013. pp. 2373-2386. DOI: 10.2118/166274-ms

[66] Dohmen T, Blangy JP, Zhang J. Microseismic depletion delineation. Interpretation. 2014;**2**(3):SG1-SG13. DOI: 10.1190/INT-2013-0164.1

[67] Cipolla C, Wolters J, McKimmy M, Miranda C, Hari-Roy S, Kechemir A, et al. Observation Lateral Project: Direct Measurement of Far-Field Drainage. In: SPE Hydraulic Fracturing Technology Conference and Exhibition. The Woodlands, Texas, USA. February 2022. DOI: 10.2118/209164-MS

[68] Jacobs T. Hess research project points to fewer Wells and passive producers that "harvest" offset fractures. Journal of Petroleum Technology. 2022;**74**(04):32-39. DOI: 10.2118/0422-0032-JPT

[69] Miranda C, Cipolla C, Murray M. Analysis of well interference tests in the Bakken shale, a fully communicated system. In: SPE/AAPG/SEG Unconventional Resources Technology Conference. Houston, Texas, USA. June 2022. DOI: 10.15530/urtec-2022-3717749

[70] Chu W, Scott K, Flumerfelt R, Chen C-C. A New Technique for Quantifying Pressure Interference in Fractured Horizontal Shale Wells. In: SPE Annual Technical Conference and Exhibition. Dallas, Texas, USA. September 2018. DOI: 10.2118/191407-MS

[71] Cipolla CL, Motiee M, Kechemir A. Integrating microseismic, geomechanics, hydraulic fracture modeling, and reservoir simulation to characterize parent well depletion and infill well performance in the Bakken. In: SPE/AAPG/SEG Unconventional Resources Technology Conference. Houston, Texas, USA. July 2018. DOI: 10.15530/urtec-2018-2899721

[72] Lorwongngam AO, Cipolla C, Gradl C, Gil Cidoncha J, Davis B. Multidisciplinary Data Gathering to Characterize Hydraulic Fracture Performance and Evaluate Well Spacing in the Bakken. SPE Hydraulic Fracturing Technology Conference and Exhibition. The Woodlands, Texas, USA. February 2019. DOI: 10.2118/194321-MS

[73] Cipolla C, Litvak M, Prasad RS, McClure M. Case history of drainage mapping and effective fracture length in the Bakken. In: SPE Hydraulic Fracturing Technology Conference and Exhibition 2020, HFTC 2020. The Woodlands,

Texas, USA. February 2020. pp. 1-43. DOI: 10.2118/199716-ms

[74] Fowler G, McClure M, Cipolla C. Making sense out of a complicated parent/child well dataset: A bakken case study. In: SPE Annual Technical Conference and Exhibition. Virtual, October 2020. pp. 1-17. DOI: 10.2118/201566-ms

[75] McClure M, Kang C, Medam S, Hewson C. ResFrac Technical Writeup. 2018. Available from: http://arxiv.org/abs/1804.02092

[76] Merzoug A, Rasouli V. Offset well design optimization using a surrogate model and metaheuristic algorithms: A Bakken case study. Eng. 2023;**4**(2):1290-1305. DOI: 10.3390/eng4020075

[77] Ling K, He J, Pei P, Wang S, Ni X. Comparisons of Biot's coefficients of bakken core samples measured by three methods. In: 50th US Rock Mechanics/Geomechanics Symposium. Vol. 2. Houston, Texas. June 2016. pp. 1603-1615

[78] Ma X, Zoback MD. Laboratory experiments simulating poroelastic stress changes associated with depletion and injection in low-porosity sedimentary rocks. Journal of Geophysical Research: Solid Earth. 2017;**122**(4):2478-2503. DOI: 10.1002/2016JB013668

[79] McClure M, Albrecht M, Bernet C, Cipolla C, Etcheverry K, Fowler G, et al. Results from a Collaborative Industry Study on Parent/Child Interactions: Bakken, Permian Basin, and Montney. In: SPE Hydraulic Fracturing Technology Conference and Exhibition. The Woodlands, Texas, USA. January 2023. DOI: 10.2118/212321-MS

[80] Lane W, Chokshi R. Considerations for optimizing artificial lift in unconventionals. In: SPE/AAPG/SEG unconventional Resources Technology Conference. Denver, Colorado, USA. August 2014. DOI: 10.15530/urtec-2014-1921823

[81] Parshall J. Challenges, Opportunities Abound for Artificial Lift. JPT. 2013. Available from: https://jpt.spe.org/challenges-opportunities-abound-artificial-lift

[82] Temizel C, Canbaz CH, Betancourt D, Ozesen A, Acar C, Krishna S, et al. A comprehensive review and optimization of artificial lift methods in Unconventionals. In: Paper presented at the SPE Annual Technical Conference and Exhibition, Virtual, October 2020. DOI: 10.2118/201692-MS

[83] Ganpule S, Cherian B, Gonzales V, Hudgens P, Reyes Aguirre P, Mata D, et al. Impact of well completion on the uncertainty in technically recoverable resource estimation in Bakken and three forks. In: Paper presented at the SPE Unconventional Resources Conference Canada, Calgary, Alberta, Canada, November 2013. 2013. DOI: 10.2118/167131-MS

[84] Eisner P, Langbauer C, Frühwirth R. Comparison of a novel finite element method for sucker rod pump downhole dynamometer card determination based on real world dynamometer cards. Upstream Oil and Gas Technology. 2022;**9**:100078. DOI: 10.1016/j.upstre.2022.100078

[85] Pankaj P, Patron KE, Lu H. Artificial lift selection and its applications for deep horizontal wells in unconventional reservoirs. In: SPE/AAPG/SEG Unconventional Resources Technology Conference. Houston, Texas, USA. July 2018. DOI: 10.15530/urtec-2018-287

[86] Patron KE, Zhang K, Xu T, Lu H, Cui S. Case study of artificial lift strategy selection and optimization

for unconventional oil Wells in the Williston Basin. In: Paper presented at the SPE Liquids-Rich Basins Conference - North America, Midland, Texas, USA, September 2018. 2018. DOI: 10.2118/191793-MS

[87] Britvar J, Williams S. Improving ESP application for unconventional wells in the Bakken. In: Paper presented at the SPE Electric Submersible Pump Symposium, April 2017, The Woodlands, Texas, USA. DOI: 10.2118/185150-MS

[88] Clark CW, Griffin L, Pearson CM, Strickland S, Resources L. The case for utilizing hydraulic jet pumps in the Bakken. In: Paper presented at the SPE Annual Technical Conference and Exhibition, Amsterdam, The Netherlands, October 2014. 2014. DOI: 10.2118/170916-MS

[89] Nickell I, Treiberg T. Autonomous control of well downtime to optimize production and cycling in sucker rod pump artificially lifted wells. In: SPE Artificial Lift Conference and Exhibition - Americas. Galveston, Texas, USA. August 2022. DOI: 10.2118/209743-MS

[90] Bestgen D, Hughes B, George J, Mullins K, Beans G, Penkala J. Novel approach to extend electrical submersible pump run times in the Williston Basin. In: Paper Presented at the SPE Annual Technical Conference and Exhibition, Virtual, October 2020. 2020. DOI: 10.2118/201474-MS

[91] Orji E, Lissanon J, Oluwole O. Sucker rod lift system optimization of an unconventional well. In: Paper presented at the SPE North America Artificial Lift Conference and Exhibition, The Woodlands, Texas, USA, October 2016. 2016. DOI: 10.2118/181242-MS

[92] Rankin R, Thibodeau M, Vincent MC, Palisch TT. Improved production and profitability achieved with superior completions in horizontal wells: A Bakken/three forks case history. In: Paper presented at the SPE Annual Technical Conference and Exhibition, September 2010, Florence, Italy. DOI: 10.2118/134595-MS

[93] Yuan G, Narvaez DA, Xue H, Nagarakanti S. Well trajectory impact on production from ESP-lifted shale Wells: A case study. In: Paper presented at the SPE Electric Submersible Pump Symposium, April 2017, The Woodlands, Texas, USA. 2017. DOI: 10.2118/185145-MS

[94] Pennel M, Hsiung J. Detecting failures and optimizing performance in artificial lift using machine learning models. In: Paper presented at the SPE Western Regional Meeting, Garden Grove, California, USA, April 2018. 2018. DOI: 10.2118/190090-MS

[95] Freeman J, Kulkarni P, Benoit R, Filipi J, Arnst B. n.d.Enabling autonomous well optimization via using IoT-enabled devices and machine learning in Bakken horizontal Wells. In: Paper presented at the SPE Artificial Lift Conference and Exhibition - Americas, The Woodlands, Texas, USA, August 2018. DOI: 10.2118/190955-MS

Chapter 8

Toward Understanding the Effect of Electromagnetic Radiation on In Situ Heavy Oil Upgrading and Recovery: Background and Advancements

Reza Gharibshahi, Naser Asadzadeh and Arezou Jafari

Abstract

Electromagnetic (EM) heating, like microwave radiation, is one of the newest and most promising thermal enhanced oil recovery (EOR) methods for producing oil from heavy oil and bitumen reservoirs. The basis of this method is reducing the viscosity of heavy oil to improve its movement toward the injection well. On the other hand, the given heat to the reservoir can, in situ, upgrade the heavy oil by cracking large molecules, reducing resin and asphaltene content, and so on. This study explained the method's basic theory, mechanism, and governing equations. The background and recent developments in this field were reviewed. It found that using additional EM absorbing materials, like magnetic nanoparticles, polar solvents, and green ionic liquids, can improve the process's efficiency. The limited field-scale applications of this method showed that this method is economically feasible and has fewer environmental challenges than conventional thermal EOR methods.

Keywords: electromagnetic heating, EOR, upgrading, viscosity, Asphaltene

1. Introduction

Heavy oil refers to a type of crude oil with a high viscosity and density and low API gravity compared to conventional crude oil [1]. It is characterized by its thick, sticky consistency [2] and is often found in reservoirs that contain a mixture of bitumen, water, and other impurities. Heavy oil is typically more challenging to extract and refine than lighter crude oil grades [3]. The production of heavy oil involves various challenges due to its physical properties. Its high viscosity makes it resistant to flow, requiring specialized techniques for extraction [4, 5]. Additionally, heavy oil often contains higher amounts of sulfur, metals, and other impurities, which can complicate refining [6]. Despite these challenges, heavy oil reserves are abundant globally and significantly meet energy demands. Many countries possess substantial heavy oil resources [7, 8], including Canada's oil sands, Venezuela's Orinoco Belt, Russia's Western Siberia region, and the United States' Bakken Formation.

Moreover, heavy oil has both economic and environmental implications [9]. From an economic standpoint, heavy oil reserves provide an opportunity for energy security and economic growth in countries with significant resources. However, heavy oil extraction and refining processes can be costly due to the need for specialized equipment and technologies. From an environmental perspective, heavy oil production can have higher greenhouse gas emissions than conventional crude oil due to the energy-intensive extraction methods involved. Refining also produces more carbon dioxide emissions per barrel than lighter crude grades.

However, efforts are being made to develop technologies that can improve extraction efficiency and reduce the environmental impact of heavy oil production. Heavy oil recovery operation is the technique and process used to extract and produce heavy crude oil from reservoirs. The recovery of heavy oil involves several methods that aim to reduce the viscosity of the crude oil, allowing it to flow more easily through wells and pipelines. Hence, some common techniques utilized in heavy oil recovery include thermal methods [5], such as steam [10] or gas [11] injection and in situ combustion [12], as well as nonthermal methods like chemical flooding [13]. Thermal methods involve heating the reservoir to reduce the viscosity of the heavy oil [14]. Steam injection is one of the most widely used thermal recovery techniques, where steam is injected into the reservoir to heat the heavy oil and improve its mobility. Besides, in situ combustion involves igniting a portion of the heavy oil in place to generate heat and create a combustion front that moves through the reservoir. On the other hand, nonthermal methods focus on reducing viscosity through chemical means. Solvent injection involves injecting solvents such as propane or butane into the reservoir to dilute the heavy oil and improve its flow properties. Chemical flooding utilizes polymers or surfactants that interact with heavy oil to reduce viscosity or alter behavior. It is worth noting that selecting a specific recovery method depends on various factors such as reservoir characteristics, economics, environmental considerations, and technological feasibility. Each technique has its advantages and limitations, requiring careful evaluation before implementation. Nevertheless, heavy oil recovery is crucial in meeting global energy demands as conventional crude reserves decline [6]. But it poses significant technical challenges due to its unique properties. Ongoing research and development efforts focus on improving recovery techniques, reducing costs, and minimizing the environmental impacts of heavy oil production.

With the gradual depletion of lighter, easy-to-extract crude oil reserves, the industry increasingly focuses on heavy oil and bitumen, which are more difficult to produce. Heavy oil recovery is a critical and challenging aspect of petroleum production. One of the main challenges of heavy oil recovery is its high viscosity, which makes it difficult to flow under normal conditions. Various thermal enhanced oil recovery methods have been employed to reduce the viscosity of heavy oil, thereby facilitating its extraction. Among these methods, electromagnetic (EM) heating has emerged as a promising technique [15, 16]. EM heating is a promising technique for heavy oil recovery, offering a solution to the challenges associated with extracting heavy oil and bitumen from reservoirs. The principle behind EM heating for heavy oil recovery is to increase the temperature of the oil reservoir using EM energy. By doing so, the viscosity of the heavy oil is reduced, enabling it to flow more easily toward production wells. This method takes advantage of the fact that certain materials can absorb EM energy and convert it into heat [17]. EM waves can penetrate the reservoir, delivering heat more evenly and efficiently than other thermal methods like steam injection. The efficiency of EM heating-assisted oil recovery is influenced by various factors, including electricity costs, well type and completion status, water salinity, crude oil composition, irradiation time, power levels, and recovery scenarios. The arrangement of wells and the methodology for implementing the EM field within the reservoir significantly impact the efficacy of the process.

Two main EM heating techniques are used for heavy oil recovery: radio frequency (RF) and microwave heating. RF heating is the most commonly employed method and operates within the frequency range of 0.5 MHz - 1 GHz [18]. The heating mechanism in RF heating is predominantly dipolar [19]. As EM waves interact with the heavy oil, polar molecules within the oil are caused to oscillate rapidly, generating frictional heat. This heat reduces the viscosity of the heavy oil, allowing it to flow more freely. One of the critical advantages of RF heating is its ability to heat the reservoir uniformly, even in reservoirs with heterogeneous properties [2]. This uniform heating contributes to a more efficient oil recovery process. It helps avoid the risk of overheating, which may damage the reservoir or decrease the quality of the recovered oil. RF heating has been successfully applied in various heavy oil reservoirs worldwide, demonstrating its effectiveness and reliability.

Microwave heating is another EM heating technique used for heavy oil recovery. It operates at higher frequencies, typically around 2.45 GHz [20]. Microwave heating can heat the reservoir faster than RF heating but has a lower penetration depth. As a result, microwave heating is more suitable for near-wellbore heating [21], where it can rapidly increase the temperature of the oil near the production well. Additionally, microwave heating offers the advantage of selectively heating certain materials within the reservoir, such as water or minerals. This selective heating capability allows for greater control over the heating process and can be beneficial in specific reservoir conditions. The choice between RF and microwave heating depends on various factors, including reservoir characteristics, recovery objectives, and operational considerations. Each technique has advantages and disadvantages, and selecting the most appropriate method should be based on a comprehensive evaluation of these factors.

One significant advantage of EM heating over other thermal EOR methods is its environmental friendliness [2]. Unlike steam injection, which requires large volumes of water, EM heating consumes minimal water. This ability makes it an attractive option, particularly in arid regions where water scarcity is a concern. EM heating does not generate combustion gases, reducing its impact on air quality. This environmental aspect is increasingly important as the industry seeks to minimize its ecological footprint and adopt more sustainable practices.

Despite its potential benefits, the widespread adoption of EM heating for heavy oil recovery still faces particular challenges. One of the primary challenges is the high energy costs associated with this method. The generation and delivery of EM waves require significant energy inputs, which can affect the economic viability of the process. Efforts are underway to develop more energy-efficient technologies and optimize the use of EM heating in heavy oil recovery [22]. Technical difficulties in designing and deploying the necessary equipment for EM heating are another challenge. The equipment must efficiently generate and transmit EM waves while withstanding the oil reservoir's harsh conditions (high temperature and pressure). Innovations in equipment design and manufacturing processes are continuously explored to overcome these technical obstacles. Another consideration is the potential interference of EM heating with other electronic systems. The generation of EM waves in the reservoir can interfere with nearby electronic equipment, such as communication systems or sensors. It is crucial to thoroughly assess and mitigate potential interference issues to ensure EM heating techniques' safe and effective implementation.

Despite the extensive endeavors undertaken in this field, it is imperative to undertake novel investigations and explore alternative approaches to surmount the challenges associated with implementing this method on an industrial scale. Therefore, as the industry continues to evolve, EM heating has the potential to play a vital role in enhancing oil recovery efficiency and sustainability. Overall, this chapter acknowledges the existing challenges within this particular field and explores the potential for oil industry researchers and practitioners to implement this methodology on a larger scale in field settings.

2. Fundamentals of the EM heating process

2.1 Basic theory

EM heating is a process that utilizes EM radiation to generate heat in a material. Richey first proposed this process in 1956 to increase oil production [15]. This heating method is based on the principles of EM waves and their interaction with matter. EM heating methods can be divided into three main categories depending on the frequency of electric current: Ohmic heating or resistance heating at low frequencies (3 to 300 kHz), induction heating at medium frequencies (300 kHz to 300 MHz), and microwave heating at high frequencies (300 MHz to 300 GHz). Fundamentally, EM waves consist of electric and magnetic fields oscillating perpendicular to each other and propagating through space. These waves can transfer energy to matter when they interact with it. When an EM wave encounters a material, it can be absorbed, transmitted, or reflected depending on its properties [23]. A material's absorption of EM waves converts the wave's energy into heat. This phenomenon occurs due to the interaction between the electric field component of the wave and charged particles within the material. The charged particles, such as electrons or ions, experience forces due to the electric field oscillations, causing them to accelerate and collide with neighboring particles. These collisions increase kinetic energy and, subsequently, an increase in temperature. The absorption of EM waves and heat generation are influenced by polar substances like water [24], asphaltene, resin, and adsorbents present in the environment [25].

Consequently, heavy oil, tar sand, and oil shale exhibit a higher capacity for absorbing EM waves [26]. Applying EM heating on tight rocks increases pore-water pressure, thereby aiding in intricate fractures within reservoir rocks. Shale can absorb EM energy more than other rock types due to its elevated dielectric constant and physical structure [27]. Generally, the efficiency of EM heating depends on several factors, including the frequency of the radiation, the properties of the material being heated, and their interaction characteristics. The frequency of radiation determines its ability to penetrate materials; higher frequencies, like microwaves, can penetrate shorter into materials compared to lower frequencies, like RF. Appropriate EM radiation absorber materials are often used for efficient heating applications. For instance, water molecules strongly absorb microwave radiation [28] due to their dipolar nature, making microwaves an effective heating source for liquids containing water. In addition to absorption, other factors such as reflection and transmission also play a role in determining how efficiently EM waves heat a material. Reflection occurs when waves bounce off a surface without being absorbed or transmitted. Transmission is waves passing through a material without being significantly absorbed or reflected. Therefore, understanding the fundamentals of EM heating allows for designing and optimizing heating systems for specific applications.

The primary purpose of the EM heating method is to convert EM energy into thermal energy, increase the temperature of the reservoir, and optimally spread the heat inside the formation to crack the heavy molecules in the crude oil and reduce its viscosity. This work makes it possible to produce more and better quality oil during EM heating in heavy oil reservoirs. Various well configurations have been proposed for this process in field operations, such as using two horizontal wells (like the SAGD process) and cyclic and continuous RF heating in a vertical well. For instance, **Figure 1** shows the implementation of EM heating in the vertical well configuration.

2.2 Mechanisms and governing equations of EM heating

EM heating finds applications in several fields due to its numerous advantages, such as industrial processes like oil recovery operations [29]. One significant advantage is its efficiency in transferring heat directly to the material being heated without needing a medium like air or water. This direct transfer minimizes energy losses and allows precise control over the heating process. Furthermore, EM heating offers rapid and uniform heating compared to conventional methods like conduction or convection. Generating heat within the material allows faster heating rates and more consistent temperature distribution. Microwave heating is a selective process. This process is very suitable for reactions that occur in the cross section of two different phases (liquid/liquid and solid/liquid). These features make this technology a suitable option for use in in situ processes. Additionally, EM heating can be easily controlled by adjusting parameters such as frequency, intensity, and duration of the applied EM field. This flexibility enables precise temperature control and reduces the risk of overheating or thermal damage to sensitive materials.

However, the mechanisms behind EM heating are based on the principles of EM induction [30] and dielectric heating [31]. EM induction is the process by which an electric current is induced in a conductor when exposed to a changing magnetic field.

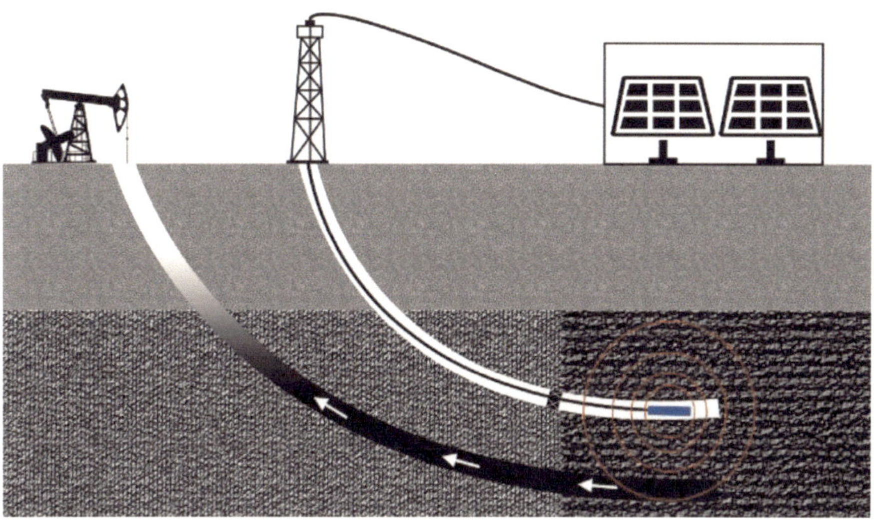

Figure 1.
EM heating implementation in the horizontal well configuration [6].

Michael Faraday first discovered this phenomenon in the early 19th century. According to Faraday's law of EM induction, the magnitude of the induced current is directly proportional to the rate at which the magnetic field changes. In EM heating, an alternating current (AC) is passed through a coil or conductor, creating a time-varying magnetic field around it. When a conductive material is placed within this changing magnetic field, eddy currents are induced within the material due to Faraday's law. These eddy currents flow in closed loops within the material and generate heat through resistive losses. The amount of heat generated by eddy currents depends on several factors, including the material's electrical conductivity, magnetic permeability, and the frequency and intensity of the applied magnetic field [6]. Materials with higher electrical conductivity, such as metals like copper or aluminum, exhibit more significant resistive losses, generating more heat.

Another mechanism involved in EM heating is dielectric heating. Dielectric materials are nonconductive substances that can store electrical energy when subjected to an electric field. When these materials are exposed to an alternating electric field, their molecules align themselves with the changing direction of the field. As these molecules continuously reorient themselves with each change in polarity of the electric field, frictional forces are generated within the dielectric material due to molecular interactions. These frictional forces result in molecular vibrations and rotations, leading to energy dissipation in heat. The amount of heat generated through dielectric heating depends on different factors, including the material's dielectric constant and loss tangent, as well as the frequency and intensity of the applied electric field. For instance, dielectric materials with higher loss tangents, like water, exhibit greater energy dissipation, generating more heat [32]. The primary relation of dielectric heating of a material in the process of EM wave radiation is as follows:

$$Q = \omega.\varepsilon_r^{''}.\varepsilon_0.E^2 \tag{1}$$

where Q is the amount of dielectric heating, ω is the frequency, ε_r" is the imaginary component of the relative permeability coefficient, ε_0 is the vacuum permeability coefficient, and E is the electric field strength (with appropriate power and intensity). So, it is clear that if the material has a more significant dielectric loss component, it can be heated more and better in an EM heating process. The process of EM heating in the context of oil recovery can be described using Maxwell's equations, heat transfer equations, and the equation of continuity. These equations together can describe the process of EM heating in oil recovery. However, it is important to note that solving these equations for a real-world application can be very complex and may require numerical methods. The actual process may also involve multiphase flow and other considerations, further complicating the equations. Overall, the governing equations of the EM heating method are summarized in **Table 1**.

Where B represents the magnetic flux density (T), ρ is the electric charge density (C/m^3), ε_0 indicates the permittivity of free space (8.85×10^{-12} F/m), μ_0 is the magnetic permeability (H/m), J represents the current density vector (A/m^2), σ is the electrical conductivity (S/m), q is related to the heat generated (W), k is the thermal conductivity (W/m.K), T represents the temperature (K), P is the power density (W/m^3), and Q indicates the heat generation rate (W). In general, EM heating mechanisms can be divided into the following three general categories:

- Magnetic Loss: It refers to the loss of different energies inside ferromagnetic compounds, such as iron and nickel. The time-varying magnetic field of

Name	Equation	References
Gauss's Law for Electric Fields	$\nabla \cdot E = \rho / \varepsilon_0$	[33]
Gauss's Law for Magnetic Fields	$\nabla \cdot B = 0$	[33]
Faraday's Law of EM Induction	$\nabla \times E = -\partial B/\partial t$	[34]
Ampere's Law with Maxwell's Addition	$\nabla \times B = \mu_0 J + \mu_0 \varepsilon_0 \partial E/\partial t$	[35]
Heat Transfer Equation - Fourier's Law	$q = -k\nabla T$	[36]
Ohm's Law	$J = \sigma E$	[37]
Power Density Equation	$P = \sigma E^2$	[38]
Heat Generation Equation	$Q = J \cdot E$	[36]

Table 1.
Summary of equations of EM heating.

microwave waves, H(t), creates magnetic spin impulses inside the magnetic material. The temperature of materials increases due to the combination of residual energy loss and eddy currents as heat inside them.

- Dipolar Polarization: It is caused by the electric field component of microwave waves, E(t), due to the continuous rotation of polar molecules to align with an oscillating magnetic field. Therefore, the temperature of the fluid increases due to the pushing, pulling, and collision of these rotating molecules with other neighboring molecules.

- Ionic Conduction: The energy dissipated as heat is due to the collision of moving charged particles (such as ions) within the solution.

3. EM absorber materials to improve heating efficiency

One of the problems of the microwave radiation process is the low penetration depth of the waves inside the formation. On the other hand, crude oil is not a good absorber for microwaves. The greater amount of radiated energy is absorbed by increasing the amount of polar components like asphaltene and resin in crude oil. Therefore, it is necessary to use EM-absorbing materials to increase the efficiency of the EM heating process and heat propagation over a larger reservoir area. EM absorbers are materials designed to absorb and dissipate EM energy. These materials find applications in various fields, including heavy oil recovery in the petroleum industry.

Nanomaterials have gained significant attention in recent years due to their unique properties at the nanoscale. When it comes to heavy oil recovery, nanomaterials can be used as EM absorbers to enhance the efficiency of the recovery process [39]. The absorption of EM waves by these nanomaterials is based on several mechanisms. One common mechanism is dielectric loss, where the nanomaterials possess high dielectric constant and loss tangent values [40]. It allows them to effectively absorb and convert the incident EM energy into heat energy. Another mechanism is magnetic loss, where the nanomaterials exhibit high magnetic permeability and loss tangent values. It enables them to absorb and dissipate EM energy through magnetic interactions. Not every nanoparticle can absorb EM waves. Meanwhile, metal nanoparticles, especially magnetic nanoparticles such as iron (Fe), nickel (Ni), and cobalt (Co), can be suitable candidates for optimal absorption of EM waves. Several factors must be considered to design effective EM absorbers for heavy oil recovery. Firstly, the nanomaterials should have a high absorption capacity over a wide range of frequencies relevant to the heavy oil recovery process [41]. It ensures they can efficiently absorb different types of EM waves generated during the recovery operations. Additionally, the nanomaterials should possess good thermal stability and chemical resistance to withstand harsh conditions encountered during heavy oil recovery processes. They should also be compatible with other components used in the recovery system. Furthermore, these nanomaterial-based EM absorbers must have a high dispersion ability in heavy oil reservoirs. It allows them for uniform distribution within the reservoir, maximizing their absorption efficiency.

Moreover, polar solvents can be used as EM absorbers to improve the efficiency of heavy oil recovery. Polar solvents are substances with a permanent dipole moment due to polar bonds or functional groups [42]. This dipole moment allows them to interact with EM fields. When exposed to an EM wave, polar solvents can absorb some EM energy and convert it into heat through molecular interactions. In heavy oil recovery, EM absorbers can enhance the heating process. Heavy oil is characterized by its high viscosity, which makes it difficult to extract from reservoirs. When the solvent is used in the EM heating process, the oil can move more easily toward the production well due to its increased dilution and reduced viscosity. EM waves can be directed toward the reservoir, where the solvent molecules absorb them using polar solvents as EM absorbers. This absorption leads to localized solvent heating and subsequently increases its temperature. The increased temperature of the solvent helps reduce the viscosity of heavy oil, making it easier to flow and recover from the reservoir. This heating process can also lead to other beneficial effects, such as reduced interfacial tension (IFT) between oil and water, improved rock surface wettability, and enhanced oil mobility within the reservoir [43]. The choice of polar solvents as EM absorbers depends on their ability to efficiently absorb EM radiation in a specific frequency range. Depending on their molecular structure and properties, different polar solvents may have varying absorption characteristics.

Besides, ionic liquids are a class of materials that consist of ions, which are charged particles and are typically liquid at or near room temperature [44]. They have gained significant attention recently due to their unique properties and potential applications in various fields, including heavy oil recovery. In heavy oil recovery, ionic liquids can act as efficient EM absorbers. EM waves are composed of electric and magnetic fields that oscillate perpendicularly. When these waves encounter a material, they can be absorbed, reflected, or transmitted depending on the material's properties. Ionic liquids possess several characteristics that make them suitable for EM absorption in heavy oil recovery processes. Firstly, they have high electrical conductivity due to charged ions [45]. This conductivity allows them to interact with EM waves effectively and absorb their energy. Secondly, ionic liquids can be tailored to have specific properties by selecting appropriate cations (positively charged ions) and anions (negatively charged ions). This tunability enables researchers to design ionic liquids with optimal EM absorption properties for heavy oil recovery applications.

Furthermore, the viscosity of ionic liquids can be adjusted by modifying their chemical structure [46]. This property is crucial in heavy oil recovery as it allows the ionic liquid to penetrate through the reservoir efficiently and improve the mobility of heavy oil. The mechanism behind EM absorption by ionic liquids involves the conversion of EM energy into thermal energy. When an EM wave interacts with an ionic liquid, it induces molecular motion within the liquid due to its high electrical conductivity. This molecular motion generates heat through frictional forces, leading to the dissipation of EM energy as thermal energy. Several benefits can be achieved by using ionic liquids as EM absorbers in heavy oil recovery processes. Firstly, the absorbed EM energy can help increase the temperature within the reservoir. This temperature rise reduces the viscosity of heavy oil and enhances its flowability, making it easier to recover. Secondly, using ionic liquids can improve the efficiency of EM heating methods, such as EM induction or microwave heating. By absorbing and converting more EM energy into heat, ionic liquids can enhance the overall heating efficiency and reduce energy losses.

4. Background and advancements in EM-heating oil upgrading and recovery

4.1 EM heating for *in-situ* upgrading of heavy oil

in situ upgrading of heavy crude oil using EM heating is a technique employed in the oil industry to enhance the recovery and processing of heavy oil reserves [22]. The in situ upgrading process begins with the drilling of wells into the heavy oil reservoir. EM heating systems are installed in these wells, typically in electrical conductors or antennas. RF energy is transmitted through these antennas into the reservoir formation, generating heat within the oil-bearing rock. The EM waves induce molecular polarization and agitation within the heavy oil, causing it to heat up. This temperature increase reduces the oil's viscosity, making it easier to flow through the reservoir and toward the production wells. The heated oil can then be pumped to the surface for further processing. One of the critical advantages of in situ upgrading using EM heating is its ability to selectively heat the oil and the reservoir rock while minimizing heat loss to the surrounding formation [47, 48]. This targeted heating helps avoid excessive energy consumption and ensures efficient oil extraction. The studies conducted in the *insitu* upgrading of crude oil using EM radiation are divided into several general categories, as shown in **Figure 2**.

Several numerical models have been presented in the EM heating process to predict fluid movement, heating area, temperature distribution, and so on. Studies have shown that the rate of heating increases with rising frequency [15]. Higher EM frequencies can overcome problems caused by medium discontinuities (fractures) that cause EM waves to propagate. The process can be controlled and adjusted to optimize the heating profile, depending on the characteristics of the reservoir. in situ upgrading using EM heating offers several potential benefits. It can increase the recovery factor of heavy oil reserves by improving oil mobility and reducing the need for costly diluents or solvents. It also enables the extraction of heavy oil that would otherwise be uneconomical to produce using conventional methods.

Furthermore, heating can facilitate partially upgrading heavy oil within the reservoir. The elevated temperatures can induce chemical reactions that lead to the decomposition of asphaltenes and the reduction of sulfur and metal content, resulting in a higher-quality oil product. Asphaltene induces alterations in the rheological properties of crude oil, leading to elevated viscosity, diminished mobility, compromised oil quality, and reduced production in reservoirs. When subjected to EM radiation, asphaltene particles within this oil variety assimilate the energy from these waves and undergo fragmentation into smaller hydrocarbon chains, thereby

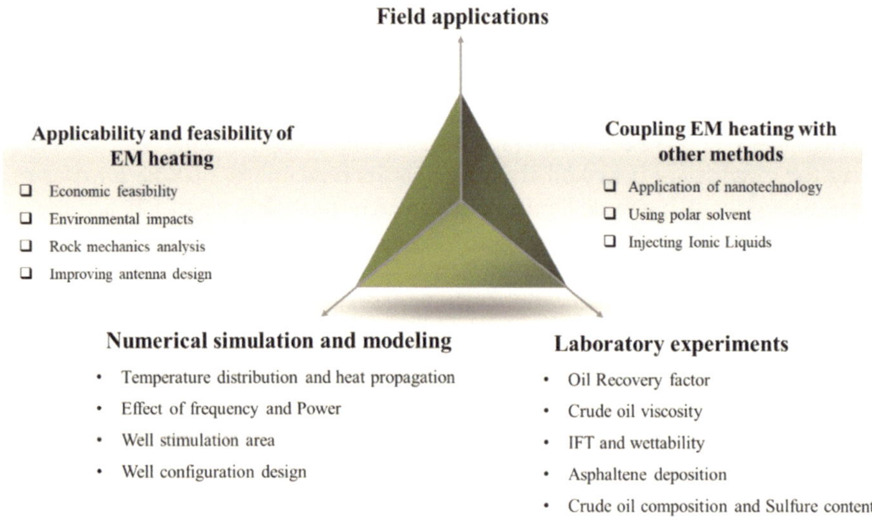

Figure 2.
Classification of conducted studies in in situ *upgrading using EM radiation.*

causing a decrease in viscosity. In this regard, it has been proved that the variations in crude oil viscosity are nonlinearly correlated with both the power of EM waves and the proportion of asphaltene in crude oil [49]. This in-site upgrading can reduce the costs of transporting and refining heavy oil and mitigate environmental concerns related to high sulfur content. However, it is essential to note that in situ upgrading using EM heating is still an emerging technology, and specific challenges limit its widespread application. These include the high power requirements, potential technical difficulties associated with installing and maintaining EM heating systems, and the need for comprehensive reservoir characterization to optimize the heating process.

4.2 Recent advantages in EM heating oil recovery

EM radiation technology utilizes the interaction between EM waves and hydrocarbon molecules to enhance oil recovery. The EM spectrum encompasses a wide range of wavelengths, from radio waves with long wavelengths to gamma rays with short wavelengths. When EM waves interact with hydrocarbon molecules, they induce molecular vibrations and rotations, increasing temperature. The process involves transmitting high-frequency EM waves into the reservoir through antennas or electrodes on or near the wellbore. These waves penetrate deep into the reservoir, where they encounter hydrocarbon molecules. As a result, energy is transferred from the EM waves to the hydrocarbons, causing them to heat up. Low-frequency waves tend to penetrate deeper but are more prone to attenuation due to absorption by water or conductive minerals present in the formation [50]. It is crucial to select frequencies that match well with the dielectric properties of both hydrocarbons and surrounding rock formations to optimize heating efficiency. It ensures that most of the energy is absorbed by hydrocarbons rather than being wasted on nonproductive regions.

A further advantage of EM heating oil recovery is its compatibility with various reservoirs. Unlike steam injection or hot water flooding, which are primarily suitable for high-permeability reservoirs, EM heating can be applied to high-permeability and low-permeability formations [51]. This versatility makes it viable for various reservoir conditions, expanding its applicability in different geographical locations. Moreover, recent advancements in EM heating oil recovery technology have focused on optimizing the design and configuration of the EM heating system. Researchers have been exploring different antenna geometries [52], such as dipole or loop antennas, to enhance energy transfer efficiency into the reservoir. Improving antenna design makes achieving higher power densities within the formation possible, resulting in faster and more effective heating. In addition to optimizing antenna design, researchers have also been investigating various methods for controlling and monitoring the temperature distribution within the reservoir during EM heating. Real-time temperature monitoring using fiber optic sensors or distributed temperature sensing (DTS) systems allows for a better understanding of the heating process. It enables adjustments to be made to optimize oil recovery [53]. This level of control and monitoring is crucial for ensuring efficient and safe operation of the EM heating system.

Another recent EM heating oil recovery advancement is integrating this technology with other enhanced oil recovery (EOR) techniques. For instance, researchers have further explored combining EM heating with chemical flooding methods, such as nanoparticles, to improve oil recovery efficiency [54, 55]. Nanoparticles exhibit a pronounced affinity for EM waves and possess notable dielectric constants [56–58]. Consequently, employing these particles as catalysts to augment heat production and expedite reaction rates represents a viable approach for enhancing oil recovery and achieving more efficient upgrades. Nevertheless, the synergistic effects of these combined techniques can lead to enhanced oil displacement from the reservoir and increased ultimate recovery. As an illustration, **Table 2** exhibits a selection of nanoparticles employed with EM radiation to enhance the oil recovery.

EM properties of nanoparticles, such as dielectric constant, loss tangent, magnetic saturation, and electrical conductivity, play an essential role in the ability of nanoparticles to absorb the energy radiated to them optimally. Nanoparticles transfer the irradiated EM energy to the surrounding environment (injection fluid and oil) through hybrid heating mechanisms, including conductive, convective, and even radiation (due to the very high temperature of the particles). Two general ways to

Nanoparticles	Effectiveness	References
ZnO	Increasing oil recovery factor	[59]
TiO_2, TiO_2-Fe_3O_4	Crude oil viscosity reduction, changing wettability	[39]
γ-Al_2O_3	Temperature rising and oil upgrading	[60]
Fe, Fe_2O_3, Fe_3O_4	Viscosity reduction	[22]
Ni, NiO	Viscosity reduction and increasing oil recovery	[56]
Carbon Nanocatalysts	Oil temperature rising and viscosity reduction	[61]
Nano Ferro	Increasing temperature and oil recovery factor	[62]
Mn_2O_3	Increasing oil recovery factor	[63]
Fe_3O_4-MWCNT	Improve EM absorption, increase oil recovery	[39]
Fe_3O_4-NiO	Crude oil viscosity reduction, changing wettability	[41]

Table 2.
Summary of nanoparticles-assisted EM radiation effect on the oil recovery process.

improve the ability of a material to absorb EM waves have been reported: one is to reduce their size to nanoscale, and the other is to hybridize them with other materials. Recently, the use of magnetic nanohybrids such as iron oxide-multiwalled carbon nanotubes (Fe_3O_4-MWCNT) as strong EM absorbers has been investigated. The hybridization of Fe_3O_4 nanoparticles with MWCNT adds the dielectric loss mechanism caused by the polarization of carbon nanotubes to the magnetic loss mechanism of microwave absorption of Fe3O4 nanoparticles. It makes the nanohybrid absorb the radiated EM energy's electric and magnetic terms. Therefore, during the process of EM radiation, they will heat up more and better and transfer it to the surrounding medium optimally.

One of the recent advances in using EM heating in oil reservoirs has been the simultaneous injection of a solvent (such as butane or pentane) under EM radiation. Using solvent causes it to dilute the crude oil, reducing viscosity and improving the movement of oil in the porous medium. It will increase the production rate of crude oil and, at the same time, reduce environmental pollution. The primary purpose of this method, which is known as Effective Solvent Extraction Incorporating EM Heating (ESEIEH), is more control and flexibility in managing the energy required in the thermal EOR production methods from shale oil and bitumen reservoirs. This method can be implemented in a wide range of conventional and unconventional reservoirs without needing a large volume of water to produce steam and with less energy. This method reduces greenhouse gases such as carbon dioxide (CO_2) emissions (generally ~50–70%). Production of a cheap solvent with high polarity, which has a high ability to absorb EM energy, can increase the efficiency of this method.

Ionic liquids are a new class of green solvents that have received much attention recently. As green solvents, these compounds have an excellent potential to overcome the disadvantages of conventional methods of increasing oil production, which should be specially investigated. Ionic liquids have advantages such as low vapor pressure, high density, high thermal and chemical stability, and the ability to synthesize various ionic and cationic components. Ionic liquids increase the amount of oil recovery by reducing the surface tension forces between oil and water and changing the wettability of the reservoir rock from oil-wet to water-wet. Due to their high polarity, these liquids have shown an excellent ability to absorb EM waves, which can prevent asphaltene precipitation.

Lastly, recent advancements in computational modeling and simulation tools have significantly contributed to developing EM-heating oil recovery [64]. Numerical simulations allow engineers and researchers to predict and optimize the performance of EM heating oil recovery systems under different reservoir conditions. These models consider reservoir heterogeneity, fluid properties, and EM field distribution, providing valuable insights into system design and operational parameters.

5. Economic justification and environmental issues

5.1 Economic feasibility

The economic feasibility of EM heating for heavy oil recovery depends on several factors, including the cost-effectiveness of the technology compared to alternative methods, the availability and cost of energy sources, and the specific conditions of the oil reservoir. It involves assessing the investment required for implementing the technology, including equipment costs, installation, and operation. The potential increase in oil production and associated revenues must also be considered. Comparative economic evaluations should be conducted to determine how EM heating performs against other extraction methods, such as steam injection or solvent-based processes. EM heating requires significant energy to generate the EM field and heat the reservoir. The availability and cost of energy sources play a crucial role in the economic feasibility of the technology. The energy can be supplied by electricity, natural gas, or other fuels. Hence, the cost of these energy sources will impact the overall operational expenses of the EM heating system.

Moreover, the proximity and accessibility of energy infrastructure to the oil field should also be considered. Also, the heavy oil deposits' geological and reservoir characteristics influence EM heating's economic feasibility. Factors such as the reservoir's depth, thickness, and permeability affect oil recovery efficiency using EM heating [2]. The presence of water or gas zones within the reservoir can also impact the performance and economic viability of the technology. Detailed reservoir studies, including core analysis and simulation modeling, are essential to assess the potential benefits and limitations of EM heating in a specific reservoir. The economic feasibility of EM heating may also depend on the scale of the project and the anticipated duration of oil production. Pilot projects may have different economic considerations than large-scale commercial operations [65]. Furthermore, the time required to recover the investment and achieve a positive return on investment (ROI) should be carefully evaluated. The longevity and sustainability of the EM heating system should be assessed to determine its economic viability over the expected project lifespan.

Most importantly, the economic feasibility of any oil recovery technology, including EM heating, is influenced by regulatory requirements and environmental considerations [15]. Compliance with environmental regulations, such as emissions control and water usage, can impact the costs of implementing and operating the EM heating system. Nevertheless, potential environmental benefits, such as reduced greenhouse gas emissions compared to other extraction methods, may also have economic implications, including potential carbon credits or incentives. It is important to note that the economic feasibility of EM heating for heavy oil recovery can vary significantly depending on the specific project and regional factors. Detailed feasibility studies and economic analyses, considering all relevant parameters and uncertainties, are necessary to determine the viability of implementing EM heating technology in a particular oil field.

5.2 Environmental impacts

EM heating requires significant energy to generate and maintain the EM field. This energy is typically obtained from fossil fuel sources, contributing to greenhouse gas emissions and climate change. The combustion of fossil fuels used to generate electricity for EM heating can release pollutants such as sulfur dioxide, nitrogen oxides, and particulate matter into the atmosphere [66]. These pollutants can contribute to air pollution and have detrimental effects on human health and ecosystems. Besides, the extraction process may involve injecting chemicals into the ground along with EM heating to aid in oil recovery. If not properly managed, these chemicals can contaminate groundwater sources, potentially affecting drinking water supplies and harming aquatic ecosystems. On the other hand, the construction and operation of EM heating facilities may require clearing land for infrastructure development, including well pads, pipelines, and power transmission lines [67]. This land disturbance can lead to habitat loss and fragmentation, impacting local flora and fauna.

In some cases, EM-heating heavy oil recovery techniques can induce seismic activity [68] due to the injection of steam or other fluids into the ground. It can potentially lead to earthquakes or ground subsidence, causing damage to infrastructure and posing risks to nearby communities. Furthermore, the heavy oil recovery process generates waste materials such as spent solvents, produced water, and solid residues. Proper management and disposal of these wastes are essential to prevent soil, water, and air contamination. However, to mitigate these environmental impacts, it is crucial to implement appropriate environmental management practices such as using renewable energy sources for electricity generation, implementing water recycling and treatment systems, employing proper waste management techniques, and conducting thorough environmental impact assessments before initiating EM-heating heavy oil recovery projects.

5.3 Challenges, future development, and opportunities

The development of EM heating for heavy oil recovery presents challenges and opportunities. Scientifically understanding these aspects is crucial for advancing the technology and realizing its potential in the oil industry. Hence, some of the critical challenges and opportunities associated with the development of EM-heating for heavy oil recovery have been described:

1. Static and Dynamic Experimental Analysis:

    - Studying the primary mechanism of the EM heating process and the kinetics of its reactions in the reservoir can help understand the behavior of the heavy oil and bitumen phases in the porous medium.

    - Finding the effect of different parameters, such as power and frequency, on oil recovery requires conducting different dynamic studies. Therefore, micromodel and core injection experiments can help to understand how the injection fluid moves in the porous media under EM radiation.

    - EM wave radiation can affect crude oil production mechanisms, such as wettability alteration and surface tension reduction, which must be carefully investigated.

    - These waves can provide the energy needed to crack large molecules (such as asphaltene). Therefore, the effect of EM radiation on the quality of crude oil and asphaltene deposition requires more studies.

    - The effect of EM wave radiation on rock mechanics, such as mechanical resistance and surface properties of reservoir rock, plays a vital role in formation damage issues.

2. Synthesize strong EM absorber materials;

    - The best type of nanoparticle that is most effective in this process is not yet known. Synthesis of hybrid nanoparticles can significantly help increase nanoparticle efficiency in the EM heating process.

- The instability of nanoparticles in harsh reservoir conditions reduces the efficiency of the heating process and damages the formation by reducing the permeability of the reservoir rock. Developing an optimal method to modify the surface of nanoparticles prevents the aggregation and agglomeration of nanoparticles and increases their efficiency in absorbing EM energy.

- The uniform dispersion of nanoparticles inside the reservoir improves the heating speed. Due to its high polarity, water can be a suitable candidate as an injection fluid. However, the role of other solvents as dispersion mediums and carriers of nanoparticles in the absorption of EM waves should be investigated.

- Salinity reduces nanofluids' stability and increases EM waves' absorption power simultaneously. Therefore, the effect of the salt type and the base fluid salinity on this process should be explored.

- Different production techniques are currently used depending on the type of nanoparticle to be synthesized (metal, metal oxide, etc.) for cost-effective synthesis of nanoparticles. The use of plant extracts, waste, and disposable or recycled materials, as well as the use of simple methods, can reduce the cost of producing nanoparticles. However, researchers and industrialists should consider conducting additional studies to propose a suitable method for producing nanoparticles on an industrial scale and in high tonnage.

3. Reservoir Characterization and Modeling:

- Accurately characterizing the reservoir properties and understanding their influence on the EM heating process is a significant challenge. Variations in the reservoir's electrical conductivity, permeability, and fluid saturation can impact heat distribution and ultimate recovery efficiency.

- Developing reliable simulation models that can capture the complex physics of EM heating in heterogeneous reservoirs is vital. Incorporating accurate reservoir data and accounting for multiphase flow behavior, heat transfer, and EM properties are essential for predictive modeling and optimization.

- The mechanism of EM heating using nanoparticles is complex. Therefore, to find the dominant mechanism and the determining stage of the heating rate, more research (especially numerical studies) should be done in the future.

4. Engineering and Design Challenges:

- EM heating requires the design and implementation of efficient and robust heating systems. Developing EM heating technologies that can withstand the harsh conditions of oil reservoirs, such as high temperatures, pressures, and corrosive environments, is a significant engineering challenge.

- Designing effective antenna systems that can generate and distribute EM fields uniformly throughout the reservoir is critical. Achieving optimal antenna placement and configuration to maximize heat transfer and minimize energy losses is a complex engineering task.

- Maintaining the integrity and insulation of the wellbore and surrounding formations during EM heating operations is essential. Addressing challenges related to wellbore sealing, electrical insulation, and potential formation damage is crucial for successful implementation.

5. Energy Efficiency and Optimization:

- Enhancing the energy efficiency of EM heating systems is an ongoing challenge. Minimizing energy losses during the generation and transmission of EM fields is critical to improve the overall efficiency of the process.

- Optimizing the power and frequency of the EM field to achieve efficient heating while minimizing energy consumption is an essential area of research. Understanding the complex interactions between the EM field and the reservoir's properties is necessary for effective optimization.

- Using additional materials injected into the reservoir can help reduce the amount of energy required. Therefore, the optimal design and synthesis of EM-absorbing materials on an industrial scale can lead to maximum absorption of radiated energy and heat distribution in a larger reservoir volume. Magnetic hybrid nanoparticles and green ionic liquids can be suitable options.

6. Scale-Up and Field Deployment:

- Scaling up EM heating from laboratory-scale experiments to field applications poses practical challenges. Ensuring the scalability and robustness of the technology while maintaining cost-effectiveness is a critical consideration.

- Field deployment of EM heating systems requires careful planning, including well configuration design, installation, and operational considerations. Addressing logistical challenges, such as power supply, equipment maintenance, and monitoring, is crucial for successful implementation.

7. Environmental and Regulatory Considerations:

- Understanding and addressing potential environmental impacts associated with EM heating, such as induced seismicity, groundwater contamination, and emissions, is essential. Developing mitigation strategies and complying with regulatory requirements are essential for the sustainable development of the technology.

- Exploring potential synergies between EM heating and environmental goals, such as carbon capture and storage (CCS) or utilizing renewable energy sources, presents opportunities for enhancing the environmental performance and acceptance of the technology.

6. Conclusion

Electromagnetic (EM) radiation is an alternative technology for the future of thermal enhanced oil recovery processes for optimal heating of heavy oil and bitumen reservoirs through different mechanisms, including magnetic loss, dipolar polarization, and ionic conduction. In this process, by converting electromagnetic energy into thermal energy, the large molecules of crude oil, such as asphaltene and resin, are broken, and the viscosity of crude oil is reduced. In this method, the quality and the amount of crude oil can be improved simultaneously. Crude oil is not an appropriate absorber for EM radiation like microwaves. Therefore, using EM absorbent materials such as magnetic nanoparticles and nanohybrids, green ionic liquids, and polar solvents to accelerate the heating process and increase the stimulated area inside the reservoirs and the efficiency of the process is contributory. Numerical and laboratory studies and a limited number of pilot-scale applications in America, Canada, and Russia have shown that this method is economically feasible. By addressing challenges and capitalizing on opportunities, EM heating has the potential to contribute to the efficient recovery of heavy oil resources, thereby enhancing energy security and reducing environmental impacts in the oil industry.

Author details

Reza Gharibshahi*, Naser Asadzadeh and Arezou Jafari*
Faculty of Chemical Engineering, Tarbiat Modares University, Tehran, Iran

*Address all correspondence to: rrezaa_gh@yahoo.com and ajafari@modares.ac.ir

References

[1] Martínez-Palou R, de Lourdes MM, Zapata-Rendón B, Mar-Juárez E, Bernal-Huicochea C, de la Cruz C-LJ, et al. Transportation of heavy and extra-heavy crude oil by pipeline: A review. Journal of Petroleum Science and Engineering. 2011;**75**(3–4):274-282. DOI: 10.1016/j.petrol.2010.11.020

[2] Sivakumar P, Krishna S, Hari S, Vij RK. Electromagnetic heating, an eco-friendly method to enhance heavy oil production: A review of recent advancements. Environmental Technology & Innovation. 2020;**20**: 101100. DOI: 10.1016/j.eti.2020.101100

[3] Zhao F, Liu Y, Lu N, Xu T, Zhu G, Wang K. A review on upgrading and viscosity reduction of heavy oil and bitumen by underground catalytic cracking. Energy Reports. 2021;**7**: 4249-4272. DOI: 10.1016/j.egyr.2021.06.094

[4] Xu ZX, Li SY, Li BF, Chen DQ, Liu ZY, Li ZM. A review of development methods and EOR technologies for carbonate reservoirs. Petroleum Science. 2020;**17**:990-1013. DOI: 10.1007/s12182-020-00467-5

[5] Panahov GM, Abbasov EM, Jiang R. The novel technology for reservoir stimulation: in situ generation of carbon dioxide for the residual oil recovery. Journal of Petroleum Exploration and Production. 2021;**11**:2009-2026. DOI: 10.1007/s13202-021-01121-5

[6] Farshadfar H, Armandi HS, Gharibshahi R, Jafari A. Simultaneous electromagnetic radiation and Nanofluid injection and their interactions in EOR operations: A comprehensive review. Journal of Magnetism and Magnetic Materials. 2023;**580**:170863. DOI: 10.1016/j.jmmm.2023.170863

[7] Caineng ZOU, Guangming ZHAI, Zhang G, Hongjun W, Zhang G, Jianzhong L, et al. Formation, distribution, potential and prediction of global conventional and unconventional hydrocarbon resources. Petroleum Exploration and Development. 2015;**42**(1): 14-28. DOI: 10.1016/S1876-3804(15)60002-7

[8] Hongjun WANG, Feng MA, Xiaoguang TONG, Zuodong L, Zhang X, Zhenzhen WU, et al. Assessment of global unconventional oil and gas resources. Petroleum Exploration and Development. 2016;**43**(6):925-940. DOI: 10.1016/S1876-3804(16)30111-2

[9] Byakagaba P, Mugagga F, Nnakayima D. The socio-economic and environmental implications of oil and gas exploration: Perspectives at the micro level in the Albertine region of Uganda. The Extractive Industries and Society. 2019;**6**(2):358-366. DOI: 10.1016/j.exis.2019.01.006

[10] Razeghi SA, Mitrovic V, Adjei MS. The influence of steam injection for enhanced oil recovery (EOR) on the quality of crude oil. Petroleum Science and Technology. 2017;**35**(13): 1334-1342. DOI: 10.1080/10916466.2017.1327970

[11] Gharibshahi R, Jafari A, Asadzadeh N. Influence of different nanoparticles on the gas injection performance in EOR operation: Parametric and CFD simulation study. The Canadian Journal of Chemical Engineering. 2023;**1**:1-14. DOI: 10.1002/cjce.25071

[12] Zhang X, Liu Q, Fan Z, Liu Q. An in situ combustion process for recovering heavy oil using scaled physical model. Journal of Petroleum Exploration and Production Technology. 2019;**9**(4):

2681-2688. DOI: 10.1007/s13202-019-0668-z

[13] Asadzadeh N, Ahmadlouydarab M, Haddad AS. Effects of temperature and nanofluid type on the oil recovery from a vertical porous media in antigravity fluid injection. Chemical Engineering Research and Design. 2023;**193**:394-408. DOI: 10.1016/j.cherd.2023.03.046

[14] Rehman MM, Meribout M. Conventional versus electrical enhanced oil recovery: A review. Journal of Petroleum Exploration and Production Technology. 2012;**2**:157-167. DOI: 10.1007/s13202-012-0034-x

[15] Bera A, Babadagli T. Status of electromagnetic heating for enhanced heavy oil/bitumen recovery and future prospects: A review. Applied energy. 2015;**151**:206-226. DOI: 10.1016/j.apenergy.2015.04.031

[16] Taheri-Shakib J, Shekarifard A, Naderi H. Experimental investigation of comparing electromagnetic and conventional heating effects on the unconventional oil (heavy oil) properties: Based on heating time and upgrading. Fuel. 2018;**228**:243-253. DOI: 10.1016/j.fuel.2018.04.141

[17] Taheri-Shakib J, Shekarifard A, Naderi H. Heavy crude oil upgrading using nanoparticles by applying electromagnetic technique. Fuel. 2018;**232**:704-711. DOI: 10.1016/j.fuel.2018.06.023

[18] Bientinesi M, Scali C, Petarca L. Radio frequency heating for oil recovery and soil remediation. IFAC-PapersOnLine. 2015;**48**(8):1198-1203. DOI: 10.1016/j.ifacol.2015.09.131

[19] Mutyala S, Fairbridge C, Paré JJ, Bélanger JM, Ng S, Hawkins R. Microwave applications to oil sands and petroleum: A review. Fuel Processing Technology. 2010;**91**(2):127-135. DOI: 10.1016/j.fuproc.2009.09.009

[20] Bientinesi M, Petarca L, Cerutti A, Bandinelli M, De Simoni M, Manotti M, et al. A radiofrequency/microwave heating method for thermal heavy oil recovery based on a novel tight-shell conceptual design. Journal of Petroleum Science and Engineering. 2013;**107**:18-30. DOI: 10.1016/j.petrol.2013.02.014

[21] Wang H, Rezaee R, Saeedi A. Preliminary study of improving reservoir quality of tight gas sands in the near wellbore region by microwave heating. Journal of Natural Gas Science and Engineering. 2016;**32**:395-406. DOI: 10.1016/j.jngse.2016.04.041

[22] Gharibshahi R, Omidkhah M, Jafari A, Mehrooz N. Parametric optimization of in-situ heavy oil upgrading using simultaneous microwave radiation and magnetic nanohybrids via Taguchi approach. Fuel. 2022;**325**:124717. DOI: 10.1016/j.fuel.2022.124717

[23] Hong J, Xu P, Xia H, Xu Z, Ni QQ. Electromagnetic interference shielding anisotropy enhanced by CFRP laminated structures. Composites Science and Technology. 2021;**203**:108616. DOI: 10.1016/j.compscitech.2020.108616

[24] Lin L, Yuan S, Chen J, Xu Z, Lu X. Removal of ammonia nitrogen in wastewater by microwave radiation. Journal of hazardous materials. 2009;**161**(2–3):1063-1068. DOI: 10.1016/j.jhazmat.2008.04.053

[25] Zhang Y, Adam M, Hart A, Wood J, Rigby SP, Robinson JP. Impact of oil composition on microwave heating behavior of heavy oils. Energy & Fuels. 2018;**32**(2):1592-1599. DOI: 10.1021/acs.energyfuels.7b03675

[26] Taheri-Shakib J, Kantzas A. A comprehensive review of microwave application on the oil shale: Prospects for shale oil production. Fuel. 2021;**305**: 121519. DOI: 10.1016/j.fuel.2021.121519

[27] Chen JH, Georgi DT, Liu HH. Electromagnetic thermal stimulation of shale reservoirs for petroleum production. Journal of Natural Gas Science and Engineering. 2018;**59**: 183-192. DOI: 10.1016/j.jngse.2018.08.029

[28] Parmar H, Asada M, Kanazawa Y, Asakuma Y, Phan CM, Pareek V, et al. Influence of microwaves on the water surface tension. Langmuir. 2014;**30**(33): 9875-9879. DOI: 10.1021/la5019218

[29] Armandi HS, Jafari A, Gharibshahi R. Nanoparticles assisted microwave radiation: Fluid-rock interactions in oil reservoirs. Petroleum Science. 2021;**18**(6):1797-1812. DOI: 10.1016/j.petsci.2021.09.002

[30] Burke HE. Electromagnetic induction. In: Burke HE, editor. Handbook of Magnetic Phenomena. Dordrecht, Netherlands: Springer; 1986. pp. 177-189. DOI: 10.1007/978-94-011-7006-2_11

[31] Leonelli C, Veronesi P, Cravotto G. Microwave-assisted extraction: An introduction to dielectric heating. In: Chemat F, Cravotto G, editors. Handbook of Microwave-Assisted Extraction for Bioactive Compounds. Food Engineering Series. New York, United States: Springer; 2012. pp. 1-14. DOI: 10.1007/978-1-4614-4830-3_1

[32] Grant E, Halstead BJ. Dielectric parameters relevant to microwave dielectric heating. Chemical Society Reviews. 1998;**27**(3):213-224. DOI: 10.1039/A827213Z

[33] Marder B. A method for incorporating Gauss' law into electromagnetic PIC codes. Journal of Computational Physics. 1987;**68**(1): 48-55. DOI: 10.1016/0021-9991(87)90043-X

[34] Galili I, Kaplan D, Lehavi Y. Teaching Faraday's law of electromagnetic induction in an introductory physics course. American Journal of Physics. 2006;**74**(4):337-343. DOI: 10.1119/1.2180283

[35] Bay F, Labbé V, Favennec Y, Chenot JL. A numerical model for induction heating processes coupling electromagnetism and thermomechanics. International journal for numerical methods in engineering. 2003;**58**(6): 839-867. DOI: 10.1002/nme.796

[36] Hossan MR, Dutta P. Effects of temperature dependent properties in electromagnetic heating. International Journal of Heat and Mass Transfer. 2012; **55**(13–14):3412-3422. DOI: 10.1016/j.ijheatmasstransfer.2012.02.072

[37] Tomassone MS, Widom A. Friction forces on charges moving outside of a conductor due to Ohm's law heating inside of a conductor. American Journal of Physics. 1997;**65**(12):1181-1183. DOI: 10.1119/1.18770

[38] Yang B, Lei H, Bi Q, Jiang J, Zhang H, Zhao Y, et al. Electromagnetic conditions in a tundish with channel type induction heating. Steel research international. 2018;**89**(10):1800145. DOI: 10.1002/srin.201800145

[39] Gharibshahi R, Omidkhah M, Jafari A, Fakhroueian Z. Hybridization of superparamagnetic Fe3O4 nanoparticles with MWCNTs and effect of surface modification on electromagnetic heating process efficiency: A microfluidics enhanced oil recovery study. Fuel. 2020;

282:118603. DOI: 10.1016/j.fuel.2020.118603

[40] Phang SW, Tadokoro M, Watanabe J, Kuramoto N. Microwave absorption behaviors of polyaniline nanocomposites containing TiO2 nanoparticles. Current Applied Physics. 2008;**8**(3–4):391-394. DOI: 10.1016/j.cap.2007.10.022

[41] Gharibshahi R, Omidkhah M, Jafari A, Fakhroueian Z. Experimental investigation of nanofluid injection assisted microwave radiation for enhanced heavy oil recovery in a micromodel system. Korean Journal of Chemical Engineering. 2022;**39**(3):562-575. DOI: 10.1007/s11814-021-0961-7

[42] Tekin N, Cebe M, Tarımcı Ç. Polarizabilities and dipole moments of benzaldehyde, benzoic acid and oxalic acid in polar and nonpolar solvents. Chemical Physics. 2004;**300**(1–3):239-246. DOI: 10.1016/j.chemphys.2004.01.022

[43] Hassan YM, Guan BH, Chuan LK, Halilu A, Adil M, Adam AA, et al. Interfacial tension and wettability of hybridized ZnOFe2O3/SiO2 based nanofluid under electromagnetic field inducement. Journal of Petroleum Science and Engineering. 2022;**211**:110184. DOI: 10.1016/j.petrol.2022.110184

[44] Lozano P, Martinez-Sanchez M. Ionic liquid ion sources: Characterization of externally wetted emitters. Journal of Colloid and Interface Science. 2005;**282**(2):415-421. DOI: 10.1016/j.jcis.2004.08.132

[45] Vila J, Gines P, Rilo E, Cabeza O, Varela L. M. Great increase of the electrical conductivity of ionic liquids in aqueous solutions. Fluid Phase Equilibria. 2006;**247**(1–2):32-39. DOI: 10.1016/j.fluid.2006.05.028

[46] Noble RD, Gin DL. Perspective on ionic liquids and ionic liquid membranes. Journal of Membrane Science. 2011;**369**(1–2):1-4. DOI: 10.1016/j.memsci.2010.11.075

[47] Adam M, Anbari H, Hart A, Wood J, Robinson JP, Rigby SP. In-situ microwave-assisted catalytic upgrading of heavy oil: Experimental validation and effect of catalyst pore structure on activity. Chemical Engineering Journal. 2021;**413**:127420. DOI: 10.1016/j.cej.2020.127420

[48] Zhu J, Yi L, Yang Z, Li X. Numerical simulation on the in situ upgrading of oil shale reservoir under microwave heating. Fuel. 2021;**287**:119553. DOI: 10.1016/j.fuel.2020.119553

[49] Gunal OG, Islam MR. Alteration of asphaltic crude rheology with electromagnetic and ultrasonic irradiation. Journal of Petroleum Science and Engineering. 2000;**26**(1–4):263-272. DOI: 10.1016/S0920-4105(00)00040-1

[50] Qi Q, Fu LY, Deng J, Cao J. Attenuation methods for quantifying gas saturation in organic-rich shale and tight gas formations. Geophysics. 2021;**86**(2):D65-D75. DOI: 10.1190/geo2020-0291.1

[51] Davletbaev A, Kovaleva L, Babadagli T. Heavy oil production by electromagnetic heating in hydraulically fractured wells. Energy & Fuels. 2014;**28**(9):5737-5744. DOI: 10.1021/ef5014264

[52] Abdulrahman MM, Meribout M. Antenna array design for enhanced oil recovery under oil reservoir constraints with experimental validation. Energy. 2014;**66**:868-880. DOI: 10.1016/j.energy.2014.01.002

[53] Nath DK, Sugianto R, Finley D. Fiber-optic distributed-temperature-

sensing technology used for reservoir monitoring in an Indonesia steam flood. SPE Drilling & Completion. 2007; **22**(02):149-156. DOI: 10.2118/97912-PA

[54] Hassan YM, Guan BH, Zaid HM, Hamza MF, Adil M, Adam AA, et al. Application of magnetic and dielectric nanofluids for electromagnetic-assistance enhanced oil recovery: A review. Crystals. 2021;**11**(2):106. DOI: 10.3390/cryst11020106

[55] Ali H, Soleimani H, Yahya N, Khodapanah L, Sabet M, BMR D, et al. Enhanced oil recovery by using electromagnetic-assisted nanofluids: A review. Journal of Molecular Liquids. 2020;**309**:113095. DOI: 10.1016/j.molliq.2020.113095

[56] Greff J, Babadagli T. Use of nano-metal particles as catalyst under electromagnetic heating for in-situ heavy oil recovery. Journal of Petroleum Science and Engineering. 2013;**112**: 258-265. DOI: 10.1016/j.petrol.2013.11.012

[57] Bera A, Babadagli T. Effect of native and injected nano-particles on the efficiency of heavy oil recovery by radio frequency electromagnetic heating. Journal of Petroleum Science and Engineering. 2017;**153**: 244-256. DOI: 10.1016/j.petrol.2017.03.051

[58] Nasri Z. Upgrading vacuum distillation residue of oil refinery using microwave irradiation. Chemical Engineering and Processing-Process Intensification. 2019;**146**:107675. DOI: 10.1016/j.cep.2019.107675

[59] Alnarabiji MS, Yahya N, Nadeem S, Adil M, Baig MK, Ghanem OB, et al. Nanofluid enhanced oil recovery using induced ZnO nanocrystals by electromagnetic energy: Viscosity increment. Fuel. 2018;**233**:632-643. DOI: 10.1016/j.fuel.2018.06.068

[60] Pourafshary P, Al FH. Enhancement of heavy oil recovery by nanoparticle/microwave application. Eurasian Chemico-Technological Journal. 2019; **21**(1):51-56. DOI: 10.18321/ectj790

[61] Li K, Hou B, Wang L, Cui Y. Application of carbon nanocatalysts in upgrading heavy crude oil assisted with microwave heating. Nano Letters. 2014; **14**(6):3002-3008. DOI: 10.1021/nl500484d

[62] Indriani E, Noor RM, Rachmat S, Munir A. Drainage of remaining heavy oil reservoir by implementing electromagnetic microwave with nano-ferro fluid injection. In: 2018 Progress in Electromagnetics Research Symposium (PIERS-Toyama); 01–04 August 2018; Toyama, Japan. New York: IEEE; 2018. pp. 1304-1309

[63] Hasani M, Jafari A. Electromagnetic field's effect on enhanced oil recovery using magnetic nanoparticles: Microfluidic experimental approach. Fuel. 2022;**307**: 121718. DOI: 10.1016/j.fuel.2021.121718

[64] Wang Z, Gao D, Fang J. Numerical simulation of RF heating heavy oil reservoir based on the coupling between electromagnetic and temperature field. Fuel. 2018;**220**:14-24. DOI: 10.1016/j.fuel.2018.02.012

[65] Mukhametshina A, Martynova E. Electromagnetic heating of heavy oil and bitumen: A review of experimental studies and field applications. Journal of Petroleum Engineering. 2013;**2013**:1-7. DOI: 10.1155/2013/476519

[66] Perera F. Pollution from fossil-fuel combustion is the leading environmental threat to global pediatric health and equity: Solutions exist. International

Journal of Environmental Research and Public Health. 2018;**15**(1):16. DOI: 10.3390/ijerph15010016

[67] Jones NF, Pejchar L, Kiesecker JM. The energy footprint: How oil, natural gas, and wind energy affect land for biodiversity and the flow of ecosystem services. Bioscience. 2015;**65**(3):290-301. DOI: 10.1093/biosci/biu224

[68] Sun Q, Retnanto A, Amani M. Seismic vibration for improved oil recovery: A comprehensive review of literature. International Journal of Hydrogen Energy. 2020;**45**(29): 14756-14778. DOI: 10.1016/j.ijhydene.2020.03.227